HOW TO FIND CHEMICAL INFORMATION

HOW TO FIND CHEMICAL INFORMATION

A Guide for Practicing Chemists, Teachers, and Students

ROBERT E. MAIZELL
Olin Corporation
Research Center
New Haven, CT.

A WILEY-INTERSCIENCE PUBLICATION

JOHN WILEY & SONS
New York • Chichester
Brisbane • Toronto

Library of Congress Cataloging in Publication Data

Maizell, Robert Edward, 1924–
How to find chemical information.

"A Wiley-Interscience book."
Includes index.
1. Chemical literature. I. Title.

QD8.5.M34 540'.7 78-23222
ISBN 0-471-56531-8

Printed in the United States of America

10 9 8 7 6 5 4

FOR MY MOTHER AND FATHER
AND FOR MONA, LIZ, AND SUE

PREFACE

One of the basic premises on which this book is written is that of change. Just as chemistry and chemical engineering change, so do the sources of information chemists and engineers use. New and improved information tools are constantly being introduced, and, concurrently, older tools become less valuable, become obsolete, or are discontinued. Accordingly, this book presents the most important and enduring of the classical tools of chemical information; the more significant newer tools; and, most importantly, the underlying methods, principles, and keys the chemist and engineer need to cope with the constantly changing array of chemical information sources and tools.

There are reasons why rapid, sometimes dramatic, change is an integral part of the chemical information scene. One is the advances and improvements made possible by computerized information handling and processing techniques. Another is the sharp escalation in publishing costs. This, when coupled with the vast amount of chemical information, has caused the demise of some older standard sources, significant changes in other tools, and numerous recent innovative attempts to provide improved approaches.

Because this book emphasizes the more enduring principles that lead to the most effective use of chemical information, the coverage of sources, methods, and tools is selective. And because most chemists and engineers are employed at some time in industry, emphasis is on approaches to meet practical needs. The equally important needs and interests of chemists and engineers in academic work (both teachers and students), and in government or independent research and development work, are also emphasized.

Whenever possible, comments are made on the pros and cons of the major sources. These comments should aid the reader in his evaluation of other sources. Additionally, estimates as to future outlook and developments are given as appropriate.

As applied to the needs of chemists in research and development and others in similar functions, this work is written in a climate reflecting the changing emphasis in programs and expenditures, which now stress: (1) the toxicological and other safety aspects, including pollution abatement and control; (2) improvement of existing products and processes; (3) the development of new products, which in many cases are built on existing strengths rather than new departures; and (4) an increasingly close relationship with the marketing function.

Information science students, teachers, and practitioners with special interest in chemistry and chemical engineering will also find material of value in this volume.

ROBERT E. MAIZELL

New Haven, Connecticut
November 1978

ACKNOWLEDGEMENTS

I am grateful to my colleagues at Olin Corporation for providing outstanding and timely insights into their specialties as well as the broad overview. Words of special thanks go to Mr. R. H. Bachman, Mr. W. M. Clarke, Dr. L. A. Krause, and Mr. D. D. Palm.

Dr. Russell J. Rowlett, Jr., Editor, Chemical Abstracts Service, contributed importantly to review of the chapter on *Chemical Abstracts*. In so doing, he also helped provide an appropriate total perspective and indirectly fine-tuned my writing emphasis and style. Mrs. Helen Lawlor, Institute for Scientific Information, was most helpful in developing a review of the innovative activities of her organization. Several managers at the United States Patent Office aided in sharpening my views on the role of patents as an information source for chemists. Dr. Brian Gore, Derwent Publications, Ltd., London, England, and Dr. Phillip Pollick, Chemical Abstracts Service, also contributed insight on patents. Dr. David R. Lide and Dr. Howard White, United States National Bureau of Standards, strengthened more fully the discussion of physical properties and related data and provided invaluable material, parts of which are briefly excerpted.

All material of a copyright nature is used here with permission of the copyright owners, to whom the author expresses his appreciation. These include the following:

Figures 5, 9, 10, 11, 14, 16, and 18 are reprinted from *Chemical Abstracts*. Material reproduced from *Chemical Abstracts* is copyrighted by the American Chemical Society and is reproduced with permission. No further copying is permitted.

Figure 6, reprinted with permission of Smithsonian Science Information Exchange, Inc., Washington, D.C.

Figure 17, reprinted from *Gmelins Handbuch der Anorganischen Chemie*, Springer-Verlag, Heidelberg. With permission.

Material beginning on page 73, with thanks to the Institute for Scientific Information (ISI), 325 Chestnut Street, Philadelphia, Pennsylvania. Other material from ISI as referred to elsewhere in this book also used with permission of ISI, including Figures 2, 3, 7, and 12 as well as appropriate textual material.

Page 108, © Verlag Chemie GmbH, *Ullmanns Encyklopaedie der Technischen Chemie*, published by Verlag Chemie Weinheim, New York. Volumes 1–6 *Thematic Section General Principles and Methodology*.

Page 109, reprinted from J. J. McKetta, *Encyclopedia of Chemical Processing and Design*, Volume descriptive material, Marcel Dekker, Inc., New York, New York.

Material beginning on page 112, with thanks to *Beilsteins Handbuch der Organischen Chemie*, Springer-Verlag, Heidelberg. With permission.

Material beginning on page 116, with thanks to *Gmelins Handbuch der Anorganischen Chemie*, Springer-Verlag, Heidelberg. With permission.

Page 136 and subsequent paragraphs on Derwent, with thanks to Derwent Publications Ltd., London, England.

Material beginning on page 195, with thanks to Dr. Bruno J. Zwolinski, Executive Director, Thermodynamics Research Center, A&M Research Foundation, P.O. Box 130, College Station, Texas 77843.

Material beginning on page 196, with thanks to Dr. Y. S. Touloukian, Director CINDAS, Purdue University, West Lafayette, Indiana.

Excerpts, pages 198–199 with permission of American Chemical Society, Washington, D.C.

Material beginning on page 225 and other material relating to Chemical Data Services, Stamford Street, London, England, used with thanks and permission.

Material beginning on page 221, 231, with thanks to John C. Dean, SRI International, Menlo Park, California.

Excerpts beginning on page 232 copyright 1976 by Chem Systems, Inc. With permission.

I am indebted to members of the John Wiley staff who helped importantly in the shaping of the manuscript and production of the book. These include Georgia Chuhay Smith, Editor; Frances Tindall, Editorial Supervisor; and Shirley Thomas, Production Supervisor.

R.E.M.

SOME CAVEATS

The terms *chemist*, *chemical engineer*, and *engineer* are meant to be used and understood as similar or related in most parts of this book. In most places the *chemist* is used as the term of choice, primarily to save space and reading time. Somewhat similar reasoning applies to the use of *he* as the arbitrarily preferred term over *she;* both terms are implied in every case.

This book represents the views and opinions of the author; it does not necessarily reflect those of his organization.

Although the material presented is aimed primarily at the United States audience, the author sincerely hopes that readers in other countries will find this volume helpful in their work and studies.

CONTENTS

FIGURES

HOW TO FIND CHEMICAL INFORMATION

1 SOME BASIC CONCEPTS

Why use the literature and other chemical information sources? This question is asked with surprising frequency by both beginning and experienced chemists and engineers. A premise of this book is that use of the literature and other information sources is key to achieving success in research and development and other functions in the chemical profession.

Effective use of information helps avoid duplicating previously reported work. This achieves savings in time and funds and avoids infringing on the proprietary rights of others. Additionally, even if there is no directly related previous work, the chemist who makes effective use of information can plan and act on a solid foundation of background data. Further, as a source of ideas or for idea development, chemical information sources and tools are invaluable fountains of inspiration and serendipity. At least one research study shows that creative chemists use the literature more than less creative chemists, although literature use does not necessarily make a person creative. (1-1)

Published sources of chemical information are not always the first choice. It is often easier and more efficient to ask a colleague in the next laboratory—especially if that person is believed to be both knowledgeable and reliable and has the desired information readily at hand. Alternatively, the chemist may find it quicker to determine needed data in the laboratory than to consult the literature, particularly for common physical properties (such as melting or boiling points) which may be conveniently measured with readily available instruments.

Nevertheless, the chemist who knows how to use chemical

information quickly and efficiently, and who has the required energy, imagination and zeal, will usually have a clear advantage over the person who either lacks these skills and qualities or is too lazy to use them.

Use of the literature and other chemical information sources is not easy. But the often significant benefits make it well worth the additional effort required.

THE SURROGATE CONCEPT IN LITERATURE USE

Some personal familiarity with—and regular use of—the journal and patent literature and other primary information sources is imperative. But as these sources continue to grow, so do specialized secondary services (such as new computer-based tools) designed to provide quicker, more effective access to the primary literature. The secondary sources are now so numerous and complex and change so rapidly that they are often best used by chemists with specialized training in information science. These persons are designated by such job titles as chemical information specialists, literature chemists, information chemists, and chemistry librarians.

Recognition is given to this specialization within the framework of the American Chemical Society by the activities of the Division of Chemical Information to which these specialties are a prime interest.

For laboratory chemists, these persons are *surrogates* who can be of considerable help and who should be used, as appropriate, to do some literature work and to advise and guide the laboratory chemist.

As noted, laboratory chemists can find surrogates valuable colleagues. But there is no substitute for firsthand familiarity with the literature. It is a mistake to rely too heavily on a surrogate for information-gathering needs.

One reason why the chemist must use the surrogate with caution is that the surrogate is usually a generalist—he cannot have the highly specialized subject expertise most laboratory chemists possess. Also, the surrogate will usually zero in on the specific topic as directed by the laboratory chemist. This means that there is a good probability that the surrogate may miss related information of importance that would be picked up by a person who specializes in the field. Also, the element of serendipity can be lost. Finally, only by some firsthand use of literature can the chemist be assured that he is keeping himself stimulated, up-to-date, and current in his own and related fields of specialization.

THE ADMINISTRATOR AND INFORMATION

If a chemical organization or department is to flourish and prosper to the full, it is important that the administrators appreciate the value of literature and other information sources and of the surrogates (literature experts) mentioned above. Literature and information use and searching are probably the most economical forms of chemical activities and can be the most productive. There is significant infusion of pertinent new information into the organization, and there are the benefits derived from cost avoidance. Support by management of the chemical information center or library is vital to the fruitful work of the chemists or engineers in the organization, be they students, professors, or industrial or governmental practioners.

2 INFORMATION FLOW AND COMMUNICATION PATTERNS IN CHEMISTRY

The chemist should be aware of the broad framework of information flow within which he works. For example, he should be aware that many chemists and other scientists obtain much of their information from colleagues by such informal means as face-to-face conversations, telephone calls, and correspondence. Informal networks (contacts) of this type are called "invisible colleges."

In addition, chemists acquire much information at national, regional, and local technical meetings sponsored by professional societies and trade associations. Primary journals, patents, review journals, books, and abstracting and indexing services are significant sources of information. The well-informed chemist is also aware of the relatively new computerized current alerting and awareness services. All of these modes are described later or referred to in this book.

The sequence of communications between chemists often follows this approximate sequence of overlapping principal steps:

1 Conduct search of literature and other information sources, including contacts with colleagues, to identify previous work, to build on a foundation of facts, and to avoid replication of what others have already done.

2 Perform laboratory work.

3 Enter data in laboratory notebooks. These are internally used, proprietary documents. Notebooks are especially important in

obtaining patent protection. Entries are best made in accordance with requirements of the organization with which the chemist is affiliated, hence the exact mode of entry and type of data required is variable. Retention of notebooks is usually governed by special records retention procedures as determined by the chemist's organization.

4 Write letters for in-house use. These are important to the inner workings of most organizations. Some locations find centralized correspondence centers useful. Others encourage personal filing systems. Letters are not a substitute for reports.

5 Write research reports and related documents for in-house use. These are proprietary internal documents in most cases. An exception would be, for example, reports based on government sponsored research that, if unclassified, are usually available to the public.

6 Submit patent applications.

7 Informally exchange results with colleagues in other organizations (face-to-face meetings, telephone calls, correspondence). Excluded from such an exchange would be information that is regarded by the chemist or his organization as confidential, proprietary, or potentially injurious to establishing a sound patent position.

8 Present results at professional society, trade association, and other technical meetings. Write and submit paper to journal editors for possible publication. In many organizations, these actions require clearance by research directors, public relations people, and legal staff to avoid premature release of information.

9 Publication of accepted papers in journals and/or issuance of patents.

10 Announcement of publication by current awareness and alerting services.

11 Abstracting and indexing of published papers and issued patents by one or more of the major abstracting and indexing services. Input of abstracting and indexing data into machine readable and computer accessible form.

12 Printing of the corresponding abstracts and indexes in full-size, microfilm, or microfiche form.

13 Use of abstracts and indexes, both printed and computer versions, by other chemists.

14 Summarization and evaluation in review articles, monographs, encyclopedias, and data evaluation centers.

3 SEARCH STRATEGY

One of the most frequently asked questions by chemists is how to conduct an effective information search—one which will yield optimum results quickly. Developing an answer to this question usually requires formulating answers to other, more specific questions. This process is helpful in developing an overall approach and philosophy to use of the chemical literature.

The first step is to formulate the goal or objective—the information being sought—as precisely as possible to save time and avoid wasted effort. The search should be delimited within those parameters that reflect the chemist's precise interest.

Some questions that the chemist should ask himself in delimiting and conducting a search include the following examples:

1 Are my goals and objectives clearly defined? Do I know what I want, why I want it, and what I will do with it when I get it?

2 What information do I already have on hand? Have I looked thoroughly at this information to see what leads this might provide and also to avoid going over the same ground?

3 How soon do I need this information? How important is it to me or to my project? Answers to these questions will help determine how much time and effort to put into the search and will reflect priorities.

4 Before looking at the literature, have I talked with colleagues in my own organization or elsewhere who might have some information (or leads to information) at their fingertips?

5 What time period do I need to cover? Must I go all the way

back to the "beginning," or would a search of the last few years suffice? Can I limit the search to the current year, at least at the outset?

6 Do I need to make a search international in scope? Or can I limit it to a specific country or geographic area where I believe most of the important work has been or is being done?

7 Can I limit the search to certain kinds of documents? For example, am I interested in patents only or in nonpatents only?

8 What specific aspect of the field, the chemical, or the chemistry am I interested in? (If I am interested in all aspects of a large field I could be taking on more than is ordinarily possible.) An examination of the latest Collective Index to *Chemical Abstracts* will give some idea of what and how much has been published on the topic.

9 What sources appear to be most fruitful in "attacking" the search and obtaining the needed information? Before beginning a search, it is important to list sources likely to be most productive. If completeness is an objective, *all* pertinent sources should be consulted; this also helps avoid bias that can come from consulting just one source.

10 How readily available are these sources to me, and do I know how to use them properly? How specific are they to the subject at hand? (I might wish to prefer a smaller, highly specialized source or service in an area of specific interest to a larger, more generalized source or service).

11 In my overall search plan, have I worked out a strategy wherein those sources that I know will take a long time to respond (such as persons I need to correspond with) will have been contacted at the outset so that I will not be delayed?

12 Before beginning a detailed search of such sources as the complete indexes to *Chemical Abstracts*, will I check selectively such potential sources of quick answers as, for example:

- **a** Desk handbooks
- **b** *Beilstein*
- **c** *Gmelin*
- **d** Review sources (e.g., review articles)

e Those complete indexes of *Chemical Abstracts* that are "on-line"

f Monographs and treatises on the subject

All of these sources are discussed in detail later in this book.

13 Will I start with the latest available source? (This is usually the best procedure rather than use an archaic source). If I use a source such as *Chemical Abstracts*, will I work "backward" in time, that is, start with the latest index and work backward, if needed, to the earliest index? This is the preferred method. After all, why go all the way back to 1907 if everything I need was published in 1976–1978, for example.

14 Have I developed a subject search plan and a flexible, iterative subject search policy? This means, first, systematically developing an array of subjects or key words that I believe are most likely to cover my areas of interest. It then means—based on experience as the search progresses—modifying, as necessary, my original array by adding or deleting search terms and tools. Some of these changes may be a result of a need for greater breadth; others may reflect a need for being more specific. Such tools as the *Index Guide* (see p. 52) help in construction of an array.

15 Do I consider all significant sides of the question under investigation? For example, in looking for data on the reaction of compound A with compound B, it is important to look under *both* compounds in the indexes if completeness is desired. Time can often be saved by first looking under the compound with the fewest number of total index entries and then, as necessary, under the other compound.

16 Should I consider attacking my problem through approaches other than subject such as author, originating organization, patent number, or molecular formula?

17 Have I asked a surrogate, such as a chemical information scientist or similar person, to review my search strategy and plans before starting in order to get the benefit of that person's expert advice?

18 Do I keep a systematic record of my progress? (Sources looked at, information found, and index entries used should be recorded in a standard laboratory notebook rather than on loose scraps of paper.)

19 Am I alert to the possibility of serendipity—of unexpectedly finding information pertinent to another of my interests or otherwise important and stimulating but not directly related to my immediate question? Am I prepared to react to and use such information if I find it?

20 Do I know when to stop? Can I recognize the point of diminishing returns? Do I know when I have all that I need? If I cannot find the information in the literature, might it be advisable to forego the information, determine it in the laboratory, or adopt some other approach?

Additional hints on search strategy will be found on pp. 89–92. Although those pages relate to on-line computer-based searching, some of the same principles could apply in other situations.

4 KEEPING UP-TO-DATE—CURRENT AWARENESS PROGRAMS

With more than 400,000 articles, patents, books, reports, and other chemical documents published annually, how can the modern chemist or engineer keep up with new materials in fields of interest? The task is difficult even for chemists and engineers who specialize in narrow fields. But fortunately, experience has shown that many persons can develop and maintain manageable programs for keeping up to date.

At the outset, the chemist must accept the fact that 100% coverage, even in a relatively limited field of interest, is not achievable in any current awareness effort. There will be pertinent material which will be identified too late or not at all.

With this important limitation in mind, the chemist's first step is to clearly define his current professional interests—areas in which he wishes to develop or maintain knowledge. It is better to state these interests in positive terms (to state what you are interested in) than in negative terms (to state what you are not interested in).

This statement is usually called a *profile.* It may include not only subjects but also authors and organizations whose activities the chemist wishes to follow because they are known to be active in specific fields.

The chemist planning a current awareness program needs to consider carefully the finite resources available to him as an individual and to his organization. He also needs to take into account other demands on resources as well as priorities.

Thus one of the most valuable assets any person has is time. Budgeting time spent in keeping up to date is essential to achieving career objectives. About five hours per week is a good rule of thumb for most chemists who have full-time jobs. That figure can easily be doubled or tripled for full-time students. Additionally, chemists embarking on new projects about which they may know little or nothing will require considerably more time than the rule of thumb, especially in the initial phases.

Further, a full-fledged literature and patent search should precede all new laboratory projects to avoid duplicating previous efforts by others. Such a search could easily take several days or even weeks, depending on the project. Chemists who are superior laboratory workers might want to have this kind of study done by a chemical information specialist (or other surrogate.) This would save time and would be an appropriate allocation of skills.

In planning current awareness efforts, the chemist also needs to consider available funds. Current awareness programs, particularly those based on computers, can become costly. Several hundred dollars per year (1977 figures) is common for a complete computer-based awareness program for a senior chemist in industry.

The key, then, is to take stock of available resources, both personal and those of one's organization, and to develop a current awareness effort consistent with those resources and appropriate to needs.

With a time-and-dollar budget clearly defined (but flexible enough to meet the unexpected), the chemist can plan with confidence. He will be better able to cope with the continuing proliferation of chemistry journals and chemical information services, many of which are highly specialized and expensive.

Each chemist needs to make a careful decision on which publications he will scan or read cover to cover regularly and for how many hours a week. The balance of the literature (except for special items identified through "Selective Dissemination of Information" programs as described on pp.

20–29) will need to be ignored. Unless this kind of decision is made, the chemist can spend his lifetime doing nothing else but read the literature.

Flexibility, however, is important too. Journals and services improve, become more expensive, deteriorate, or collapse, and new ones are introduced. Personal and organization interests, budgets, and policies change. Accordingly, reevaluation of the plan every few months is imperative.

READING CLUBS

Teams of chemists working on large projects, especially in industry, sometimes establish "reading clubs" or similar arrangements. Each person in the team agrees to accept responsibility for covering a given segment or type of literature. The results are then shared with colleagues. This is a delicate arrangement and does not function well unless there is a close working relationship and/or a clear, enforced directive from management.

PERSONAL PLAN OF ACTION

In developing a plan of action for keeping up to date, the chemist will find it helpful to visualize a triangle as shown in Figure 1.

The base of the triangle consists of general "newsy" (but important) material with which most chemists and engineers should maintain some awareness. The middle part of the triangle consists of relatively broad areas of chemistry within which there may be some specialization. The apex represents specific materials closely allied to unique individual interests and on which most current awareness time will be spent.

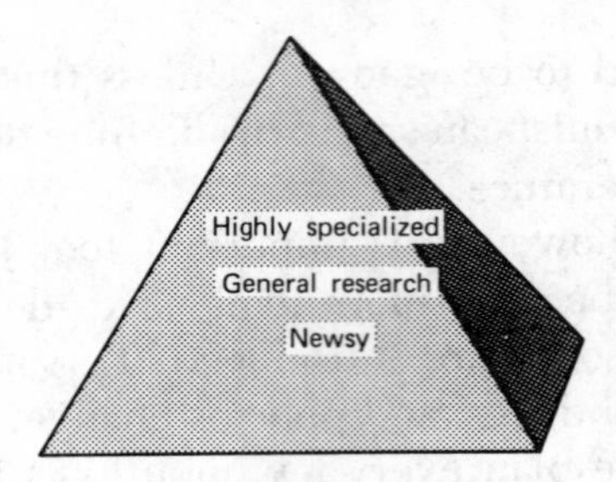

Figure 1 The current awareness triangle.

Base of the Triangle

The general material of most widespread interest is reported in chemical "news" magazines. The chemist who wants to be well informed should systematically scan "news" magazines such as:

1 *Chemical Engineering* (biweekly publication of McGraw-Hill, Inc.)

2 *Chemical and Engineering News* (weekly publication of the American Chemical Society)

3 *Chemical Marketing Reporter* (weekly publication of Schnell Publishing Co.)

4 *Chemical Week* (weekly publication of McGraw-Hill, Inc.)

5 *Chemistry and Industry* (a twice-monthly publication of the British Society of Chemistry and Industry which contains news, research, and other features)

6 *European Chemical News* (weekly publication of IPC Industrial Press Ltd.)

Any two or three of the above will keep the chemist or engineer informed on most important new developments in chemistry and chemical engineering of general interest.

Although there is some overlap in coverage among these

publications, each has unique features. For example, contributions to *Chemistry and Industry* include "Highlights from Current Literature," a monthly commentary on some of the most important and interesting recently published papers in scientific journals; there are also some original research and other papers of considerable value. *Chemical Week* and *Chemical Marketing Reporter* publish buyer's guides to chemical products. *Chemical Engineering* publishes semiannual inventories of new processes and plant construction or expansion plans. The McGraw-Hill publications and *European Chemical News* are distinguished by their speed and excellent coverage of fast-breaking, important stories and their willingness to evaluate the significance and implications of what they believe is important.

Most of the chemical "news" publications are oriented primarily toward the industrial or business aspects of chemistry. But the boundary between this kind of chemistry and the more scientific aspects is blurred. Each kind spills over into and has implications for the other. The chemist who wants to further his career will develop a balanced current awareness program and will keep informed on both the scientific and business aspects, although he will usually concentrate on one or the other.

Middle Part of the Triangle

A balanced reading program should contain two or three general research publications within which there may be some specialization and that can be skimmed selectively. Examples include:

1 *Angewandte Chemie—International Edition in English*. A top-quality monthly publication that includes important, fast-breaking developments, abstracts of important work published elsewhere, and some original research papers.

2 *Science.* Almost all branches of science are covered in this weekly, although emphasis is on the biological aspects. Important breakthroughs are frequently announced here first. *Nature* is an equally good choice.

3 One or more of the American Chemical Society's 18 primary journals.

4 Other, including commercial and trade magazines.

In cases of uncertainty, the chemist should call on the local chemistry librarian or chemical information specialist for advice in making a selection. Teachers and more experienced colleagues can also help "fine-tune" choices to achieve a manageable portfolio.

The following examples also belong in the middle part of the triangle. Reading time and costs usually permit a selection of only *one* of these examples for *highly selective* skimming:

1 *Current Contents, Physical and Chemical Sciences Edition*, published by the Institute for Scientific Information, 325 Chestnut Street, Philadelphia, PA 19106. This weekly service reproduces tables of contents of more than 700 of the most important research journals (as well as many books) in chemistry and related sciences. The service permits the chemist to quickly scan contents of many journals of potential interest. If a pertinent article is noted, the chemist can consult the full original in a local chemistry library, or can order copies using the so-called Original Article Tear Sheet service (OATS®). More details about OATS are on page 38. Disadvantages of *Current Contents* include lack of immediate access to full abstracts and unreliability of many titles as "pointers" about what is pertinent. These same disadvantages apply to most services based primarily on titles.

Chemical Titles is a similar publication issued every two weeks by Chemical Abstracts Service. Approximately 700 journals of the world's most important chemically oriented journals are covered. Each article is indexed by author and keyword with a reference to the journal in which it appears. Every article selected for coverage in *Chemical Titles* is subsequently reported in *Chemical Abstracts.* Orientation is more strongly chemical than in any other comparable United States publication.

2 *ACS Single Article Announcement (SAA)* permits the chemist to skim through contents pages of all 18 primary journals. Material of interest can be ordered directly from ACS on the form provided. Because of the narrow scope of this service, its utility is limited.

3 *Technical Survey,* published weekly by Predicasts, Inc., 11001 Cedar Ave., Cleveland, OH 44106, provides summaries of journal articles and news developments in the technology of chemistry and many other fields. Its most valuable feature is breadth of coverage, but it is relatively slow.

4 One of the Section Groupings (see pp. 66–68) of *Chemical Abstracts* provides an excellent mechanism for keeping informed in broad or specialized areas by appropriate scanning. Available groupings include:

- **a** Applied chemistry and chemical engineering
- **b** Biochemistry
- **c** Macromolecular chemistry
- **d** Organic chemistry
- **e** Physical and analytical chemistry

Because many sections are large, it is more feasible to skim quickly (to identify material of special interest) than to read exhaustively. The Section Groupings have the unique advantages of offering broader coverage than any other comparable source, full abstracts, and keyword indexes—all at a reasonable cost.

5 Scanning of one or more sections of the Derwent *Central Patents Index* or *World Patents Index* is often appropriate (see pp. 136–143 for more detail). This highly regarded tool offers comprehensive, rapid coverage of patents for most important industrial nations. Chemists and engineers working in industry find this tool of special value, since many important industrial developments are first reported in patents.

The chemist should consider selecting one of the above tools based on the best possible advice available from teachers, colleagues, and chemical information specialists or chemistry librarians.

Apex of the Triangle

The most vital part of any current awareness program is the apex of the triangle as depicted in Figure 1. This part represents the specific areas in which the chemist wants to keep as up-to-date as possible and that most closely correspond with his statement of interests or profile.

If there is a journal or other publication that zeroes in on this profile, this is one approach. Unfortunately, in many fields of chemistry there are several journals in fields of specific interest, and some of these are often too thick, no matter how specialized. Special consideration should be given to so-called letters journals such as *Tetrahedron Letters*, in which material is more likely to appear quickly than in the large, so-called omnibus journals. A totally unique journal—the only one of its kind in the field—would be the prime candidate, although such journals are relatively rare.

Also falling within the apex are computer-based tools targeted specifically at current awareness. These tools are based on input of a statement of interests (profile) matched by computer versus the total volume of input of chemical interest received by a computerized information center. (See pp. 95–99.) When a "match" or "hit" occurs, this is automatically recognized by the computer, printed, and mailed to the persons whose profiles are represented. This service is usually referred to as *Selective Dissemination of Information (SDI)*.

One SDI approach uses so-called *standard interest profiles* or *macroprofiles*. These are profiles selected by an outside organization to meet the common interests of a broad group of chemists.

A good example is ASCATOPICS ® which is sold by the Institute for Scientific Information, Philadelphia, PA. Over 460 topics are available. Topics of special interest to chemists are shown in Figure 2.

Entries in the report look like the example shown in Figure 3. Results are mailed to subscribers at a cost of $100 per year. Users receive weekly computer-based reports containing complete bibliographic information for new items published on topics they select.

Principal advantages include speed of issue and availability of the previously mentioned OATS service for convenience in getting copies of full original documents.

The list given in Figure 2, which is revised and updated annually, is reasonably representative of what was available in 1977.

ASCATOPICS, however, is limited to the approximately 5000 publications covered by the Institute for Scientific Information. Also, patents, which are especially important to industrial research workers, are not included. In an attempt to identify pertinent material from ASCATOPICS, the chemist is limited to the information supplied in the computer printout; there are no abstracts in these reports.

Macroprofiles based on *Chemical Abstracts (CA)* have been available since 1971 from the United Kingdom Chemical Information Service (UKCIS). Established in 1969, UKCIS, a directorate of the Chemical Society of London, is located at the University of Nottingham. Presently (1977) available Macroprofiles are listed in Figure 4. These current awareness bulletins, issued every other week, are produced by searching the computer-readable tapes, known as *Chemical Abstracts Condensates*. The search profiles used are refereed by experts in the field. The output contains the complete title and other bibliographic data, including the citation to *CA*. Abstracts are *not* provided in the computer-based output; the user must scan the printed *CA* for these. From this innovative macro-

CHEMISTRY

1. Aerosols
2. Alkaloids—Isolation, Characterization, Synthesis
3. Aromaticity—Alicycles, Heterocycles & Non-Benzenoid
4. Boron, Aluminum, Group 3 Elements Chemistry
5. Catalytic Reactions
6. Combustion
7. Detection & Identification of Narcotic Drugs & Poisons
8. Electrochemistry
9. Ion Exchange Research & Technology
10. Liquid Crystals
11. Luminescence
12. Manganese Chemistry
13. Organo-Silicon Compounds
14. Organic Photochemistry (Including Photolysis)
15. Peptide Synthesis
16. Photoionization
17. Polymers—Fibers & Films
18. Polymers—Metallo-Organic
19. Polymers—Preparation of New Monomers & Polymers
20. Polymers—Properties & Technology
21. Radiation Chemistry
22. Rhenium, Palladium, Iridium, Platinum & Rhodium Chemistry
23. Selenium & Tellurium Chemistry
24. Structure—Activity Relationships
25. Surface Properties

THEORETICAL CHEMISTRY

26. Catalysis

27. Chemical Reaction Mechanisms & Kinetics
28. Colloid & Surface Chemistry
29. Conformational Analysis of Small Molecules
30. Molecular Orbital Calculations
31. Optical Activity—Resolution, Racemization, ORD, CD
32. Solid State Chemistry

ANALYTIC CHEMISTRY

33. Chromatography—Affinity
34. Chromatography—Gas & GLC
35. Chromatography—Gel Permeation & Ion Exchange
36. Chromatography—High Speed Liquid
37. Chromatography—Liquid
38. Chromatography—TLC & Paper
39. Fluorescence
40. Mass Spectrometry
41. Specific Ion Electrodes
42. Spectroscopic Techniques—Atomic Absorption, Moessbauer, X-Ray
43. Spectroscopic Techniques—IR, UV & Raman
44. Spectroscopic Techniques—NMR, ESR
45. X-Ray Fluorescence Analysis

Figure 2 Selected ASCATOPICS ®

profile concept evolved the valuable service described in the following paragraph.

A new tool, available from Chemical Abstracts Service, Columbus, OH 43210, is *CA SELECTS*. This provides complete *Chemical Abstracts* abstracts and bibliographic citations. More than 70 fields are now covered in this service,

WEEKLY LITERATURE ALERTING SERVICE

A Service of ISI®

Institute for Scientific Information®
325 Chestnut Street, Philadelphia, Pennsylvania 19106 USA
Tel. (215) 923-3300. Cable: SCINFO, TELEX: 84-5305

ASCATOPICS
PHOTOLYSIS IN
ORGANIC CHEMISTRY

REPORT FOR 03 FEB 78 PAGE 1

MODEL FOR SEPARATION OF SPATIAL AND TEMPORAL INFORMATION IN VISUAL-SYSTEM
GAFNI H ZEEVI YY
BIOL CYBERN 28(2): 73-82,1977 51 REFS
THESE ITEMS IN THIS PROFILE WERE CITED:
OSTER G SCI AM 222 82 70
-----> CHECK TO ORDER TEAR SHEETS ----->( ) #EG158
H GAFNI, TECHNION ISRAEL INST TECHNOL,FAC ELECT ENGN,DEPT ELECT ENGN, HAIFA 32000, ISRAEL

PHOTO FINE-STRUCTURE OF PHOTO-KINETIC SYSTEMS IN DINOBRYON-CYLINDRICUM VAR ALPINUM (CHRYSOPHYCEAE)
KRISTIAN.J WALNE PL
BR PHYCOL J 12(4): 329-341,1977 * REFS
-----> CHECK TO ORDER TEAR SHEETS ----->( ) #EF938
J KRISTIAN., UNIV TENNESSEE,DEPT BOT, KNOXVILLE, TN 37916

ULTRAVIOLET-ABSORPTION SPECTRUM OF HYDROGEN-PEROXIDE VAPOR
MOLINA LT SCHINKE SD MOLINA MJ
GEOPHYS R L 4(12): 580-582,1977 18 REFS
THESE ITEMS IN THIS PROFILE WERE CITED:
LEVY H ADV PHOTOCHEM 9 369 74
CALVERT JG PHOTOCHEMISTRY 200 67
CALVERT JG PHOTOCHEMISTRY 200 67
VOLMAN DH ADVANCES PHOTOCHEMIS 1 43 63
-----> CHECK TO ORDER TEAR SHEETS ----->( ) #EF977
LT MOLINA, UNIV CALIF IRVINE,DEPT CHEM, IRVINE, CA 92717

REGRESSION-ANALYSIS OF NONSTATIONARY DISCHARGES IN NEURONS - ADAPTATION IN ELECTRO-SENSORY AFFERENT OF DOGFISH
BROMM B TAGMAT AT
BIOL CYBERN 28(1): 41-49,1977 31 REFS
THESE ITEMS IN THIS PROFILE WERE CITED:
MURRAY RW J PHYSIOL LOND 145 1 59
MURRAY RW J PHYSIOL LOND 180 592 65
MURRAY RW HDB SENSORY PHYSIOLO 3 74
-----> CHECK TO ORDER TEAR SHEETS ----->( ) #EG023
B BROMM, UNIV HAMBURG,KRANKENHAUS EPPENDORF, DEPT NEUROPHYSIOL, D-2000 HAMBURG 20, FED REP GER

OBTAIN ANY ARTICLE LISTED ABOVE WITH *OATS*™ ORIGINAL ARTICLE TEAR SHEET SERVICE
see *OATS* instructions on reverse side of this printout

Figure 3 Example of ASCATOPICS output—excerpts from

WEEKLY LITERATURE ALERTING SERVICE

A Service of ISI®
Institute for Scientific Information®
325 Chestnut Street, Philadelphia, Pennsylvania 19106 USA
Tel. (215) 923-3300. Cable: SCINFO, TELEX: 84-5305

ASCATOPICS
PHOTOLYSIS IN
ORGANIC CHEMISTRY

REPORT FOR 03 FEB 78 PAGE 2

INFLUENCE OF SENSITIZATION OF ELECTRON-TRANSFER THROUGH INTERFACE ZINC OXIDE ELECTROLYTE BY SALT ADDITIONS
BODE U HAUFFE K
J ELCHEM SO 125(1): 51-58,1978 47 REFS
THESE ITEMS IN THIS PROFILE WERE CITED:
MEMMING R PHOTOCHEM PHOTOBIOL 16 325 72
TRIBUTSCH H PHOTOCHEM PHOTOBIOL 14 95 71
VOGELMANN E PHOTOCHEM PHOTOBIOL 24 595 76
TRIBUTSCH H PHOTOCHEM PHOTOBIOL 16 261 72
BONNEAU R PHOTOCHEM PHOTOBIOL 25 129 77
-----> CHECK TO ORDER TEAR SHEETS ----->( ) #EG030
U BODE, UNIV GOTTINGEN, INST PHYS CHEM,
D-3400 GOTTINGEN, FED REP GER

RADIATION INDUC/
NI3SI SURFACE-FILM FORMATION CAUSED BY RADIATION-INDUCED SEGREGATION
POTTER DI REHN LE OKAMOTO PR WIEDERSI.H
SCRIP METAL 11(12):1095-1099,1977 10 REFS
-----> CHECK TO ORDER TEAR SHEETS ----->( ) #EG101
DI POTTER, ARGONNE NATL LAB,DIV MAT SCI,
ARGONNE, IL 60439

PHOTOCHEM
(GE) PHOTOCHEMOTHERAPY WITH 8-METHOXYPSORALEN AND UVA .4. SIMPLE DEVICE FOR TREATMENT OF PSORIASIS AND PUSTULOSIS PALMARIS AND PLANTARIS
MEFFERT H METZ D THORMANN T SONNICHS.N
DERMATOL M 163(12): 959-965,1977 * REFS
-----> CHECK TO ORDER TEAR SHEETS ----->( ) #EG330
H MEFFERT, HUMBOLDT UNIV,HAUT KLIN,
DDR-104 BERLIN, GER DEM REP

UV REACTION/
(GE) REACTIONS SUPPRESSED BY TOPICAL CORTICOSTEROIDS IN UV ERYTHEMA
JACOBI H KADNER H PINZER B
DERMATOL M 163(12): 970-974,1977 * REFS
-----> CHECK TO ORDER TEAR SHEETS ----->( ) #EG330
H JACOBI, MED AKAD CARL GUSTAV CARUS,HAUT KLIN,
DDR-8019 DRESDEN, GER DEM REP

PHOTOEMIS PHOTOEXCI
(FR) SPATIAL-DISTRIBUTION OF PHOTOEXCITED ELECTRONS GOLD-SILICON INTERNAL PHOTOEMISSION STRUCTURE
SEILLER JM MINN S
THIN SOL FI 47(3): 261-269,1977 13 REFS
-----> CHECK TO ORDER TEAR SHEETS ----->( ) #EG336
JM SEILLER, UNIV NANTES,UER PHYS,PHYS SOLIDE
LAB, F-44037 NANTES, FRANCE

FLASH VAPOR-PHASE PYROLYSIS OF THIOPHENE 1,1-DIOXIDES
VANTILBO.WJ PLOMP R
REC TR CHIM 96(11): 282-286,1977 44 REFS
THESE ITEMS IN THIS PROFILE WERE CITED:
DEMAYO P MOLECULAR REARRANGEM 427 63
-----> CHECK TO ORDER TEAR SHEETS ----->( ) #EG467
WJ VANTILBO., SHELL RES BV,KONINKLIJKE SHELL
LAB, AMSTERDAM, NETHERLANDS

ETHYLENE-OXYGEN METATHESIS
KIM L RALEY JH BELL CS
REC TR CHIM 96(11):M136-M137,1977 8 REFS
THESE ITEMS IN THIS PROFILE WERE CITED:
KEARNS DR J AM CHEM SOC 916554 69
FOOTE CS ACCOUNTS CHEM RES 1 104 68
-----> CHECK TO ORDER TEAR SHEETS ----->( ) #EG467

OBTAIN ANY ARTICLE LISTED ABOVE WITH *OATS*℠ ORIGINAL ARTICLE TEAR SHEET SERVICE
see *OATS* instructions on reverse side of this printout

the full report for the time period covered.

Engine Exhaust
Environmental Pollution
Gaseous Waste Treatment
Liquid Waste Treatment
Pollution Monitoring
Recovery & Recycling of Waste
Solid and Radioactive Waste
Trace Metal Analysis

Fungicides
Herbicides
Insecticides
Prostaglandins
Steroids (Chemical Aspects)
Steroids (Biochemical Aspects)

Analytical Electrochemistry
Moessbauer Studies

Colloid Chemistry (Applied Aspects)
Colloid Chemistry (Physico Chemical Aspects)
Natural Gas
Petroleum Chemistry and Technology

Metallo Enzymes and Metallo Co-Enzymes
Organo Transition Metal Complexes
Organo Fluorine Chemistry
S-Heterocycles

Computers in Chemistry
Liquid Crystals
Photobiology
Semiconducting Materials

Figure 4* *Macroprofiles available from the United Kingdom Chemical Information Service

which was introduced in 1976. Current prices for each field are in the range of $50 per year for 26 issues. Advantages include comprehensiveness and specific focus. A disadvantage is that the service is not as fast as some of those previously mentioned.

Other organizations, too numerous to mention here, also offer standard interest profile service. The principal advantages of the macroprofile concept are usually given as low cost, speed, and, sometimes, breadth of coverage. The principal disadvantage is that the coverage may not focus sufficiently on material of specific interest to an individual chemist. This can result in too much nonpertinent material. Macroprofiles may also lack the flexibility needed by some chemists who are interested in more than one field or whose interests vary sharply from time to time.

Customized SDI, based on a profile specifically designed to meet the needs of an individual chemist or a group of chemists, is usually considered more desirable than the macroprofile approach. Customizing costs more, but it is more likely to yield pertinent material and can be varied to meet changing interests. This service is available from organizations such as those listed in Figure 16. If the chemist's organization has a large computer center, he may find that this service is available to him locally. The chemist can elect to have his profile run against a variety of sources or so-called data banks. But for most individuals, *Chemical Abstracts* will be the source of choice. Typical SDI output is shown in Figure 5.

Another option is the *ACS Customized Article Service*. Under this plan ACS members can receive the following three items:

- *The ACS Single Article Announcement Service (SAA)*
- The *Journal of Medicinal Chemistry* or *Environmental Science and Technology* or *Biochemistry*
- Microfiche copies of pertinent articles selected from *SAA*

This service is experimental and will be continued or expanded, depending on response.

The supposedly ideal goal of distribution to chemists and other scientists of complete copies of only those articles of specific interest has been talked and written about for dec-

```
CA VOL. 086   NO. 14   SEC. 067   ABSTRACT NO. 096578

       REACTIONS OF CHLORAMINEPENTAAMMINEIRIDIUM(III) ION.  EVIDENCE
       FOR A COORDINATED NITRENE INTERMEDIATE

       JOHNSON, EDWARD D.    BASOLO, FRED

       INORG. CHEM., 1977, 16(3), J, 554-8,  (ENG)     CODEN-INOCAJ
       LOCATION OF WORK-DEP. CHEM., NORTHWEST. UNIV., EVANSTON, ILL.

       IRIDIUM COMPLEX REACTION KINETICS
       AMMINE IRIDIUM REACTION KINETICS
       CHLORAMINE IRIDIUM REACTION KINETICS
       IODIDE REON IRIDIUM COMPLEX
       HYDROLYSIS IRIDIUM CHLORAMINE COMPLEX
       BROMIDE REACTION IRIDIUM CHLORAMINE

CA VOL. 086   NO. 14   SEC. 068   ABSTRACT NO. 096853

       EXCESS FREE VOLUME OF BINARY LIQUID MIXTURES

       PRAKASH, SHEO    PRASAD, NARAYANI    PRAKASH, OM

       INDIAN J. PHYS., 1976, 50(9), J, 801-6,  (ENG)     CODEN-IJPYAS
       LOCATION OF WORK-CHEM. DEP., ALLAHABAD UNIV., ALLAHABAD, INDIA

       EXCESS VOL BINARY LIQ
       BENZENE BINARY MIXT EXCESS VOL
       ETHANOL BINARY MIXT EXCESS VOL
       ISOPROPANOL BINARY MIXT EXCESS VOL
       HEPTANE BINARY MIXT EXCESS VOL
       HEXANE BINARY MIXT EXCESS VOL
       XYLENE BINARY MIXT EXCESS MOL
       CHLOROMETHANE BINARY MIXT EXCESS VOL
       TOLUENE BINARY MIXT EXCESS VOL

CA VOL. 086   NO. 20   SEC. 078   ABSTRACT NO. 149829

       THERMAL STUDIES ON RARE EARTH METAL ION CHELATES OF
       8-HYDROXYQUINOLINE, 8-HYDROXYQUINALDINE AND THEIR DERIVATIVES

       CHANG, TIAO-HSU    YANG, TSONG-JEN    YEN, MEI-WAN

       J. CHIN. CHEM. SOC. (TAIPEI), 1976, 23(4), J, 181-7,  (ENG)
       CODEN-JCCTAC    LOCATION OF WORK-CHEM. RES. CENT., NATL. TAIWAN
       UNIV., TAIPEI, TAIWAN

       THERMAL STABILITY RARE EARTH QUINOLINOLATE
       QUINOLINOL CHLORO RARE EARTH STABILITY
       HYDROXYQUINALDINE CHLORO RARE EARTH STABILITY
       QUINALDINOL CHLORO RARE EARTH STABILITY
       CHLOROQUINOLINOL RARE EARTH STABILITY
       ERBIUM QUINOLINOL CHLORO THERMAL STABILITY
       LANTHANUM QUINOLINOL CHLORO THERMAL STABILITY
       PRASEODYMIUM QUINOLINOL CHLORO THERMAL STABILITY
       NEODYMIUM QUINOLINOL CHLORO THERMAL STABILITY
       SAMARIUM QUINOLINOL CHLORO THERMAL STABILITY
       GADOLINIUM QUINOLINOL CHLORO THERMAL STABILITY
       DYSPROSIUM QUINOLINOL CHLORO THERMAL STABILITY
```

Figure 5 Typical SDI output based on **Chemical Abstracts.**

ades. The ACS experiment described above is one small step in this direction. Such factors as chemist response, economics, technology, and copyright legislation will help decide whether this goal can or should be achieved on a large scale. It is not likely in the foreseeable future.

MEETINGS

As part of his current awareness program, the chemist needs to allocate time for attending professional society and trade association meetings. These meetings offer the opportunity to listen to and meet persons at the leading edge of research and development and to obtain information far before it is published. Further, the opportunity to meet informally with colleagues in the same field of interest is invaluable.

Many chemists and engineers prefer to attend more specialized smaller meetings such as Gordon Research Conferences.

In contrast, huge "omnibus" national meetings, which attempt to cover all aspects of chemistry or chemical engineering, can be too broad in scope, the quantity of papers can be overwhelming, and the number of persons attending can dilute opportunities for personal contacts as well as for good question-and-answer sessions. Attending large national meetings can be useful only if there is careful preplanning, for example, by making plans to attend several symposia on specialized topics of clear interest to the individual.

How many meetings per year can or should a chemist or engineer attend?

Like reading journals, chemists could easily spend a lifetime attending meetings. Many organizations will send their key people to an average of one meeting each year, but this can vary widely, depending on meeting location, duration, and fees. The chemist should select the meetings he wants to attend as much in advance as feasible—preferably several months.

Future developments in "teleconferencing," closed-circuit television, and wider use of audio cassettes and videotapes may make it possible for more chemists to hear more important papers selectively and promptly without traveling to distant meeting sites. For best results, this approach should have a "live" two-way communications link for questions and answers, if at all possible.

Getting Copies of Meeting Papers

Because no one can attend all meetings, what can a chemist or engineer do when he sees announcements of papers of interest but when neither he nor anyone else from his organization can attend?

Some actions to get copies of meeting papers can be attempted, but results are often uncertain.

Reliance on ultimately seeing a published version of a meeting paper is risky at best. Only about half of all meeting papers are published according to some estimates. Those published often undergo extensive revision after oral presentation and may therefore vary significantly in context from the original version. The time lag between oral presentation and formal publication can be substantial—from several months to well over a year.

American Chemical Society (ACS) meeting papers and those of the American Institute of Chemical Engineers (AIChE) are best accessed initially through the collections of abstracts of meeting papers published prior to most national and regional meetings. In the case of ACS, an index to the abstracts, based on words in the titles, speeds up identification of pertinent material.

Because meeting papers are not usually subject to rigorous review, and for other reasons, abstracts of them leave much to be desired. Chemists interested in learning more about the contents of a meeting paper can write the author and request a preprint. These are readily available, however, only for an estimated 50% or less of all meeting papers. Some suggestions on contacting authors for copies of their papers can be found on pp. 37–39.

For several divisions of ACS, obtaining meeting papers is simplified because these divisions publish preprints or reprints before or shortly after national meetings. At this writing these divisions include:

1 Chemical Marketing and Economics
2 Environmental Chemistry
3 Fuel Chemistry
4 Organic Coatings and Plastics Chemistry
5 Petroleum Chemistry, Inc.
6 Polymer Chemistry, Inc.

Additionally, meeting paper photocopy services are maintained by the Agricultural and Food Chemistry and by the Rubber Chemistry Divisions. Whether publication practices of these divisions will be continued, and whether other divisions will adopt similar practices, are points worth following closely.

For AIChE papers, microfiche copies are often available. Additionally, several weeks after the meeting, photocopies are usually available from the Engineering Societies Library, 345 East 47th Street, New York, NY 10017.

Another source is the *World Meetings* series compiled by the World Meetings Information Center in Chestnut Hill, MA and marketed by Macmillan, New York, NY.

As an aid to deciding which meetings to monitor and possibly attend, the chemical news magazines and professional society news publications publish lists on a regular basis. Additionally, there are publications that list future meetings of technical, scientific, medical, and management organizations and universities, for example, *Scientific Meetings*.

Conference Papers Index is an index of over 500,000 papers presented at scientific and technical meetings worldwide from 1973 to the present. It is available in both printed form and on-line. The publisher is Data Courier, Inc., 620 South Fifth St., Louisville, KY 40202.

Meeting paper preprints that are not published elsewhere are among the many kinds of documents covered by *Chemical Abstracts*. See p. 50 for a broad policy statement on what is covered by *Chemical Abstracts*.

The recently announced *Index to Scientific and Technical Proceedings*, published by the Institute for Scientific Information, Philadelphia, PA, is expected to help chemists and others locate conference papers published in book or journal form.

The tools just mentioned, and others that either exist now or will amost inevitably be developed in the future, are an encouraging indication that chemists should be able to get a much better "handle" on meeting and conference papers than ever before.

RESEARCH IN PROGRESS; DISSERTATIONS

Identification of research in progress can be used by chemists and engineers to:

- Avoid unwarranted duplication of research effort and expenditure.
- Locate possible sources of support for research on a specific topic.
- Identify leads to the published literature or participants for symposia.
- Obtain information to support grant or contract proposals.

- Stimulate new ideas for research planning or innovations in experimental techniques.
- Acquire source data for technological forecasting and development.
- Survey broad areas of research to identify trends and patterns or reveal gaps in overall efforts.
- Learn about current work of a specific research investigator, organization, or organizational unit.

Information about research underway is frequently difficult to obtain, since most investigators are understandably reluctant to allow premature disclosure of results. There are, however, sources that can help identify the scope of some of the work in progress.

The leading such source in the United States is the Smithsonian Science Information Exchange, Inc. (SSIE), 1730 M Street, N.W., Washington, DC 20036. This organization maintains and constantly updates a data base of information on more than 200,000 ongoing and recently completed projects in basic and applied research in the life and physical sciences, including chemistry and chemical engineering. Project information is gathered from over 1000 supporting federal government agencies, state and local governments, nonprofit organizations, and colleges and universities. This information is indexed by a staff of scientists and engineers and entered into a computerized data base for storage and retrieval. The basic record (and output) is the *Notice of Research Project*. A sample is shown in Figure 6.

The content of most of the *Notices* is "indicative"; that is, the reader gets some idea of what the work is about but not results or hard data. For these, the user of SSIE needs to contact the investigator whose name and address is usually given in full on the *Notice*.

Information can be obtained from the data base in a variety of formats. The *SSIE Science Newsletter* contains an-

FORM APPROVED
BOB NO. 105 R0002
EXPIRES 11/76

SMITHSONIAN SCIENCE INFORMATION EXCHANGE

Room 300 • 1730 M Street, N.W. • Washington, D.C. • 20036

Telephone (202) 381-4211 • Telex 89495

SSIE NUMBER: GSP-7105-7

NOTICE OF RESEARCH PROJECT

01

SUPPORTING ORGANIZATION:
U.S. National Science Foundation
Div. of Chemistry
1800 G St. N.W.
Washington, District of Columbia 20550

SUPPORTING ORGANIZATION NUMBER(S): CHE77-02185

PROJECT TITLE:
INTERACTIONS OF COMPLEX SOLUTE SPECIES WITH CHARGED INTERFACES

INVESTIGATOR(S): E MATIJEVIC

DEPARTMENT/SPECIALTY: CHEMISTRY

PERFORMING ORGANIZATION:
CLARKSON COLLEGE OF TECHNOLOGY
SCHOOL OF ARTS & SCIENCES
51 MAIN ST.
POTSDAM, NEW YORK 13676

PERIOD FOR THIS NRP: 6/77 TO 11/78
FY77 FUNDS $40,000

PROJECT SUMMARY:

This work continues a comprehensive study of chemical aspects of interfacial phenomena with special emphasis on (a) formation of colloidal dispersions consisting of particles uniform in size and shape, (b) stability of colloidal dispersions in the presence of various complex solutes, (c) particle adhesion. The program under (a) includes the preparation of monodispersed sols of a variety of metals, having particles exceedingly uniform in size, but of different shapes. It is endeavored to elucidate the chemical complexes which act as precursors in the embryonation of such monodispersed sols and to explain chemical mechanisms of formation and the mechanisms of their nucleation and growth. The program under (b) consists of several projects dealing with (1) adsorption and desorption of mixtures of hydrolyzed metal ions from lyophobic surfaces and from silica; (2) the interactions of polymers and polyelectrolytes with the defined sols; and (3) adsorption of amino acids and proteins on the colloidal particles of known characteristics. The program under (c) will be dealing with a quantitative study of parameters responsible for particle adhesion and detachment substrates of various compositions will be used (glass, metals, plastics, etc.) onto which well defined particles will be investigated as a function of the composition of rinse solutions (pH, ionic strength, the nature of electrolytes, chelating agents, surfactants, etc.). This is a renewal of Grant No., CHE-7409409.

SI-SSIE 76

Figure 6 Notice of research project, Smithsonian Science Information Exchange.

nouncements in one convenient retrieval form—research information packages on topics of high current interest. These packages consist of groups of project notices compiled and screened according to their scientific content.

The Exchange also offers custom searches of its data base, tailored by staff scientists to meet individual needs. Monthly or quarterly current awareness and on-line services (see Chapter 8) are also available.

All SSIE services are subject to a fee based on such factors as scope of the search, number of research projects, and frequency of up-dating.

SSIE covers primarily United States Government funded work. Coverage of work in other countries is far from complete at this time, although efforts are underway to improve this situation. Even for United States government work, 100% coverage cannot be expected because of secrecy regulations, proprietary considerations, and other limitations.

Doctoral dissertations offer the chemist another way of obtaining information as soon as possible and also can contain some information that may never be published in any other form.

The chemist can sometimes borrow dissertations through his local librarian if he knows the university where the work was done. But this procedure is often time consuming and, further, no loan copies are available in many cases.

A better approach is to start with the *Comprehensive Dissertation Index (CDI).* This permits searching for pertinent dissertations by subject, author, title, and university. This service, which is computer based and available on-line, is a product of University Microfilms International (UMI), 300 North Zeeb Road, Ann Arbor, Michigan 48103, or 18 Bedford Row, London, WCIR 4EJ, England. *CDI* helps locate the abstracts and/or the dissertations, most of which are available from UMI for a fee.

The abstracts are published by UMI in *American Doctoral*

Dissertations (ADD) or *Dissertation Abstracts International (DAI)*.

The program is based on agreements between individual degree-granting institutions and University Microfilms International. Most American universities now participate in this file, which goes back to 1861. In addition to United States dissertations, many from Canada and an increasing number from other countries are included.

Chemical engineers and some chemists should also be interested in the listing of recent Ph.D. dissertations that appears annually in the January issue of *Chemical Engineering Progress*. This list is categorized by subject.

5 HOW TO GET ACCESS TO ARTICLES, BOOKS, PATENTS, AND OTHER DOCUMENTS QUICKLY AND EFFICIENTLY

Many chemists report that one of the most frustrating experiences they encounter in their professional work is gaining quick access to needed articles, patents, books, and other documents not available in a local chemistry library.

The problems of obtaining access to or copies of documents quickly and efficiently have been brought home to thousands of chemists and engineers by their use of new automatic and other current alerting services (see Chapter 4). Some of these services are so fast that they list many materials before they are received by the local chemistry library. Some materials listed are so "exotic" that obtaining copies becomes a challenging, sometimes formidable task, since they are often not available locally.

Difficulty in document access is most likely to be experienced by chemists who lack ready access to strong chemistry libraries. It is also more likely to occur when the document sought is from less frequently used sources, such as journals from remote countries, material written in unfamiliar languages, or from material that is older or out of print (see section on translations, pp. 43–47).

Fortunately, there are steps the informed chemist or engineer can take to get much of what is needed. What are some of these options and solutions?

One traditional way of obtaining copies is writing authors for reprints. The first step is to be certain the reference and address are correct; this is simply a matter of proofreading in many cases. Beyond this, it is helpful to supply as much incentive as possible to encourage authors to fulfill requests for reprints quickly.

A politely worded, typed letter is helpful, preferably with indication of why the reprint desired is of interest. Even better is reciprocity, as for example, enclosing a reprint of one's own work in the same or related fields. This can help establish an ongoing rapport which may result in access to information prior to publication, depending on how close a professional bond can be established. It is also desirable to enclose a self-addressed, stamped reply envelope and/or to use a tool such as Request-A-Print® (see Figure 7).

Some authors have an antipathy to answering correspondence, especially requests for reprints. In this event, a telegram or phone call may be necessary to get what is wanted.

Alternatively, the chemist can use the Original Article Tear Sheet service (OATS) of the Institute for Scientific Information (ISI), 325 Chestnut St., Philadelphia, PA 19106. This important service covers over 5000 journals. Tearsheets are said to be provided for nearly any article covered by ISI. Orders can be placed by mail, phone, telex, and even "on-line" (see Chapter 8). So-called 24-hour response service is reportedly available on request. The OATS approach has many advantages and apparently avoids any potential copyright problems (see p. 40). Their service is best, however, only for those relatively recent publications covered by ISI; it is far from all-inclusive.

A somewhat similar service is provided by the *ACS Single Article Announcement* published by the American Chemical Society, Washington, DC 20036. This service covers only the 18 primary ACS journals. A related service available from ACS is its "Timely Tearsheets Service."

Another alternative, especially for less recent material, is

How to use Request-A-Print®

Cards conform to International Postal Card size regulations

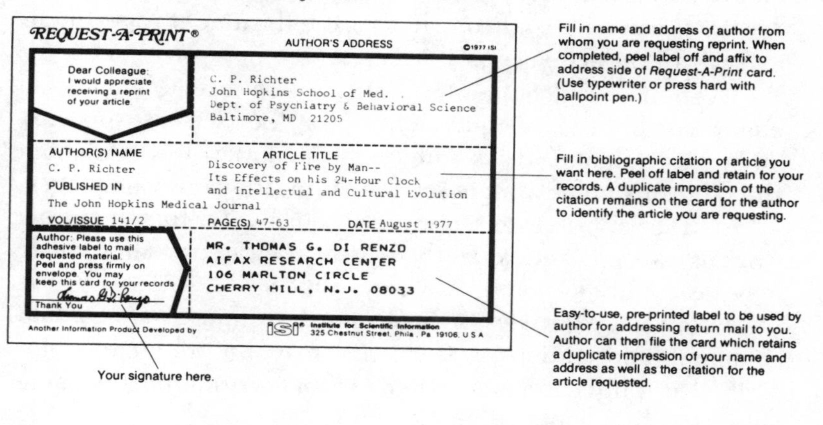
REQUEST-A-PRINT® — AUTHOR'S ADDRESS — ©1977 ISI

Dear Colleague:
I would appreciate receiving a reprint of your article.

C. P. Richter
John Hopkins School of Med.
Dept. of Psychiatry & Behavioral Science
Baltimore, MD 21205

AUTHOR(S) NAME
C. P. Richter

ARTICLE TITLE
Discovery of Fire by Man--
Its Effects on his 24-Hour Clock
and Intellectual and Cultural Evolution

PUBLISHED IN
The John Hopkins Medical Journal

VOL/ISSUE 141/2 PAGE(S) 47-63 DATE August 1977

Author: Please use this adhesive label to mail requested material. Peel and press firmly on envelope. You may keep this card for your records.
Thank You

MR. THOMAS G. DI RENZO
AIFAX RESEARCH CENTER
106 MARLTON CIRCLE
CHERRY HILL, N.J. 08033

Another Information Product Developed by ISI® Institute for Scientific Information 325 Chestnut Street, Phila., Pa 19106, U.S.A.

Fill in name and address of author from whom you are requesting reprint. When completed, peel label off and affix to address side of *Request-A-Print* card. (Use typewriter or press hard with ballpoint pen.)

Fill in bibliographic citation of article you want here. Peel off label and retain for your records. A duplicate impression of the citation remains on the card for the author to identify the article you are requesting.

Easy-to-use, pre-printed label to be used by author for addressing return mail to you. Author can then file the card which retains a duplicate impression of your name and address as well as the citation for the article requested.

Your signature here.

Request-A-Print... the efficient system for making your reprint requests.

Saves time. Gets results.

If you've done more than your share of writing and waiting, use this faster, easier, and more productive way to request reprints. Simply complete the handy order card to get your personalized *Request-A-Print* cards.

Enough writing and waiting for me! I want results!

Please send me______thousand personalized *Request-A-Print* cards. ($85 per thousand: U.S.A., Canada and Mexico. $100 per thousand: all other locations; includes delivery.)

The following information should appear on my *Request-A-Print* mailing labels:

Name & Title

Organization/Department

Street Address

City, State, Zip, Country

(Note: Limit of 30 characters per address element—spaces and punctuation count as characters. Use abbreviations as required.)

©1977 ISI A-16

Rates effective for orders received between January 1 and December 31, 1978.

ISI® **Institute for Scientific Information®**

Figure 7 Example of Request-A-Print.

for the chemist to locate one or more abstracts of the document. This can be done by a search of the abstracting services. Abstracts can help confirm whether the document is pertinent and worth ordering.

Even if the document sought is not available in local chemistry libraries, "equivalents" may be (see also pp. 45–46 on equivalents). It is common for an author to write about the same or similar work in several publications and languages.

Another option is to see if the author has written about his work as part of a series. In that case, an earlier or later paper may be readily available and may suffice.

Also, review articles or books on the subject may contain sufficient information or the work may be adequately described in publications of other authors writing on the same subject. See Chapter 9 on Reviews.

HELP AVAILABLE FROM CHEMISTRY LIBRARIANS

Chemistry librarians have a wide variety of tools for locating documents, for example, *CASSI* (5-1). This tool shows which libraries in the United States and abroad receive publications cited by *Chemical Abstracts Service* since 1907. *CASSI* also includes references to journal literature cited by *Chemisches Zentralblatt* from 1830 to 1940 and in *Beilsteins Handbuch der Organischen Chemie* through 1965 (see p. 77 and p. 112). Other information useful in locating and obtaining document copies is included.

A common library practice has been to obtain a photocopy of a needed document from local or other libraries. The new United States copyright law went into full effect in January, 1978. What impact this law will have on photocopy practices should become clear after a short time.

Another action a librarian can take, especially for books, is to borrow from another library, using the familiar procedure known as interlibrary loan. Librarians can also purchase copies of journals and other documents.

Also, librarians can help chemists determine whether "equivalents" or translations are available (see pp. 45–46).

PATENTS

Patents are not subject to copyright laws, but they may be more difficult to obtain than articles and other documents, since many chemists (and the librarians who serve them) are usually less familiar with all details of patent systems, especially non-United States patent systems.

Writing to the inventor for patent copies is not a favored approach. Instead, copies of patents (or of equivalents) can be obtained from a variety of other sources.

The chemist should first consider, however, that an article may have been written about the subject matter of the patent by the inventor or someone else in his organization. That article could supplant the need for the patent and may contain even more useful information. But articles are not a 1:1 substitute for patents; they lack the legal content and significance of patents.

Sources of patents include public and university libraries which maintain numerical sets of United States patents. See Figure 8 for a list of these libraries.

The most complete patent library in the United States is at the U.S. Patent Office, Arlington, Virginia (see p. 133).

Many libraries in industrial and other chemical research laboratories have extensive files of chemical patents in microfilm or printed form.

The chemist can also obtain copies of patents from government patent offices or from private organizations that specialize in procurement of patents.

Albany, N.Y.	University of State of New York
Atlanta, Ga.	Georgia Tech Library[a]
Birmingham, Ala.[b]	Public Library
Boston, Mass.	Public Library
Buffalo, N.Y.	Buffalo and Erie County Public Library
Chicago, Ill.	Public Library
Cincinnati, Ohio	Public Library
Cleveland, Ohio	Public Library
Columbus, Ohio	Ohio State University Library
Detroit, Mich.	Public Library
Houston, Tex.[b]	Rice University
Kansas City, Mo.	Linda Hall Library[a]
Los Angeles, Calif.	Public Library
Madison, Wis.	State Historical Society of Wisconsin
Milwaukee, Wis.	Public Library
Newark, N.J.	Public Library
New York, N.Y.	Public Library
Philadelphia, Pa.	Franklin Institute
Pittsburgh, Pa.	Carnegie Library
Providence, R.I.	Public Library
Raleigh, N.C.[b]	North Carolina State University
Seattle, Wash.[b]	University of Washington
St. Louis, Mo.	Public Library
Stillwater, Okla.	Oklahoma A.&M. College Library
Sunnyvale, Calif.	Public Library[c]
Toledo, Ohio	Public Library

[a]Collection incomplete. [b]Recent additions to this list. [c]Arranged by subject matter, collection dates from January 2, 1962.

***Figure 8** Locations of libraries that have copies of United States patents arranged in numerical order in addition to files at U.S. Patent Office, Arlington, VA.*

United States patents can be obtained for 50¢ each (1977 price) from the Commissioner of Patents and Trademarks, Washington, D.C. 20231.

Addresses of foreign patent offices, along with prices for copies of patents, are listed in the *Introduction* to Semi-Annual Volumes of *Chemical Abstracts* (found in the first issue).

Most government patent offices, although they charge relatively low prices, cannot match the delivery speed and more personalized service of commercial procurement organizations. Examples of such organizations include:

IFI Plenum/Data Co., 2001 Jefferson Davis Highway, Arlington, VA 22202

Air-Mail Patent Service, Box 2232, Arlington, VA 22202

Derwent Publications, Ltd., 128 Theobalds Road, London WC1X 8RP, England

TRANSLATIONS

One of the most vexing information problems facing chemists who are natives of the United States, Canada, the United Kingdom and other English-speaking nations is that of coping with interesting material published in a foreign language. It is a particularly troublesome problem in the United States, where relatively few chemists are multilingual.

This is still important, despite the fact that about 60% of the world's chemical literature is now published in English. This figure includes publications from countries in which some publications appear totally or partially in English, even though this is not the native language of that country. Journals published in Czechoslovakia, Japan, Sweden, and West Germany are among countries where this is sometimes the case. Examples include:

Acta Chemica Scandinavica
Angewandte Chemie (International Edition)
Bulletin of the Chemical Society of Japan
Collection of Czechoslovak Chemical Communications
Denki Kagaku (The Electrochemical Society of Japan)
Japan Chemical Week
Svensk Papperstidning

What are some options open to the chemist or engineer who learns about intriguing research published in an unfamiliar foreign language?

There are several alternative courses of action which can often help resolve the apparent dilemma relatively quickly and at minimal cost. Some of these involve working closely with a chemical information specialist or research librarian. These actions include:

1 Locate and use a suitable abstract. This may be found in an abstracting service or sometimes in the original journal. (Many foreign-language journals publish abstracts in several languages, one of which is usually English). Sometimes the abstract indicates that the material is not worth pursuing further—that it is not as pertinent as the title or other preliminary information indicated. Conversely, the abstract may suggest that the publication is of interest and worth delving into further. The abstract may contain sufficient data to satisfy immediate interest and need, but it is not wise to rely on the abstract alone. The preferred route, if the abstract indicates that the publication is of interest, is to try to look at the material in a language the chemist can understand. How to achieve this is described in some of the following paragraphs.

2 Locate and use an already available translation.

a Some journals are translated into English cover-to-cover, especially Russian-language publications, e.g.:

(**1**) *Biochemistry (Biokhimiya)*

(**2**) *Bulletin of the Academy of Sciences of the USSR: Division of*

Chemical Science (Izvestiya Akademii Nauk SSSR, Ser. Khimicheskaya)

(3) *Chemistry and Technology of Fuels and Oils (Khimiya i Tekhnologiya Topliv i Masel)*

(4) *Colloid Journal of the USSR (Kolloidnyi Zhurnal)*

(5) *Inorganic Materials (Izvestiya Akademii Nauk SSSR, Neorganischeskie Materialy)*

(6) *Journal of Analytical Chemistry of the USSR (Zhurnal Analiticheskoi Khimii)*

(7) *Journal of Applied Chemistry of the USSR (Zhurnal Prikladnoi Khimii)*

(8) *Journal of Structural Chemistry (Zhurnal Strukturnoi Khimii)*

(9) *Kinetics and Catalysis (Kinetika i Kataliz)*

(10) *Russian Chemical Reviews (Uspekhi Khimii)*

(11) *Russian Journal of Inorganic Chemistry (Zhurnal Neorganicheskoi Khimii)*

(12) *Russian Journal of Physical Chemistry (Zhurnal Fizicheskoi Khimii)*

(13) *Soviet Chemical Industry (Khimicheskaya Promyshlennost)*

b Ask the chemistry librarian to check with one of the so-called translation clearing houses to see if a translation already exists. This may be the case with material of widespread interest or importance. In the United States, one major source is the National Translations Center, John Crerar Library, 35 West 33rd St., Chicago, Illinois 60616. If the center does not have the translation, their staff may be able to suggest other sources.

3 Identify "equivalent" journal articles. As previously noted, the same author may publish on the same research in journals issued in several different countries and languages. Even authors whose native tongue is English sometimes publish first in foreign-language publications and subsequently publish the same or similar material in English. The best professional society and other publications will not accept outright double publication, but it occurs in journals with less rigid standards and in some trade publications on the applied aspects of chemistry. Identification of such duplicates is straightforward; the chemist merely needs to check the author indexes to *Chemical Abstracts* and related publications.

4 Locate review articles or books on the subject which adequately describe the work of interest. See Chapter 9 on Reviews.

5 Locate "equivalent" or corresponding patents published in English (see Chapter 11 on Patents). Identification of corresponding patents can be done with concordances such as those of *Chemical Abstracts*, Derwent, IFI, and the International Patent Documentation Center. These concordances provide information about "families" of closely related patents applied for (or issued in) different countries.

6 Forego a translation done by others and instead work with those parts of the publication that can be handled with a limited amount of dictionary look-up. Examples of well-regarded dictionaries are cited in References (5–2). This approach involves scanning for chemical equations (which require no translation) and for such parts as tables, graphs, and abstracts which may require use of a dictionary for only a few key words. (In some cases, journals in other languages will have already translated synopses or abstracts and captions into English.) It is surprising how often this simple approach will provide results at minimum cost and with acceptable time delay.

7 Identify a person within the chemist's own organization who can provide an on-the-spot oral translation. The advantage of this approach is that questions and answers between chemist and translator can readily be exchanged on points that require amplification. Additionally, an oral translation session can be limited to parts of the publication most pertinent—key sections within the text that the chemist can identify to the translator in the course of their session. Such in-house translations are usually the only way to handle confidential or proprietary materials.

8 Obtain a custom-made translation as a last resort. These can be done by commerical translation houses or by free-lance translators. Quality and accuracy of such translations vary widely. The work should be done by a translator who has a proven reputation for top quality work. The translator must have knowledge of the languages to be translated "to" and "from," and must also have adequate training as a chemist or engineer. Those organizations that translate cover-to-cover versions of chemical journals are good sources for custom-made translations or can make recommendations. Translation rates vary, depending on such factors as language, type of document, and speed of delivery required. The current commercial rates for translation from "common" languages such as German are about $25 per 1000 English words.

OTHER OPTIONS AND FUTURE OUTLOOK

It may be necessary for the chemist to personally visit a chemistry library in another city if he wants to obtain needed information as completely and quickly as possible. Prior arrangements should be made through the local chemistry librarian to increase chances of success on such visits. Some of the stronger chemistry libraries in the United States include:

1 Chemists Club Library, 52 East 41st Street, New York, N.Y. 10017

2 Engineering Societies Library, 345 East 47th Street, New York, N.Y. 10017

3 John Crerar Library, 35 West 33rd Street, Chicago, Ill. 60616

4 Library of Congress, Washington, D.C. 20541

5 Linda Hall Library, 5109 Cherry Street, Kansas City, Mo. 64110

6 New York Public Library, Science and Technology Division, 5th Avenue and 42nd Street, New York, NY 10008

7 University of California, Berkley, CA 94720

8 University of Houston, Cullen Boulevard, Houston, TX 77004

Libraries of comparable strength are located in Canada, major European nations, and Japan.

No discussion of document sources would be complete

without mention of the U.S. National Technical Information Service (NTIS), 5285 Port Royal Road, Springfield, VA 22161. Unclassified federally funded research reports and other documents are a prime emphasis of this large and important facility. Recently, plans were formulated for a photocopying service or network which will apparently embrace many important scientific and technical journals. NTIS publishes weekly abstract bulletins on its acquisitions, compiles searches in areas of importance, and has much of its files available for on-line searching.

For hard-to-get articles and other documents published in other countries, especially in the East Bloc, the U.S. State Department in Washington may be able to help.

Many chemists belong to organizations that have laboratories, plants, or other representatives and contacts abroad. Persons at these overseas locations are resources who can be called on for aid in obtaining copies of documents.

There has been discussion in the chemical community about development of a national chemistry library for the United States, either a single location or on a shared basis, at several locations. Implementation of this proposal would help chemists get materials they need more quickly.

An interesting speculation is that the present journal publication system will eventually be augmented, or perhaps even replaced, by selective, computer-based distribution of copies of articles and other documents on a large scale. Copyright payments could be built into the cost of such a system. The concept seems unlikely, at least for the next decade.

One thing is certain: more attention will be paid to document access at the local, regional, and national levels. Chemists will want to watch for developments that will help them get the documents they need as expeditiously as possible.

6 THE CHEMICAL ABSTRACTS SERVICE

By far the most heavily used and valuable information tool in chemistry, and the one best known to most chemists and chemical engineers, is *Chemical Abstracts* (*CA*), the chief product of the Chemical Abstracts Service (CAS). CAS is a division of the American Chemical Society (ACS).

Initiated in 1907, *CA* has been located since 1909 in Columbus, OH, on land in or near the Ohio State University. The current director of CAS is D. B. Baker, the associate director for planning and development is Dr. Fred A. Tate, and the editor is Dr. Russell J. Rowlett, Jr. The total full-time staff includes about 1200 persons, and the budget (1977) is about $30 million.

CA is so important—and perhaps so complex—that the ACS offers an audio/visual course on its use (6-1). CAS offers a full-day seminar on the use of *CA* and an explanatory workbook (6-2). Further, in the introductions to *CA*, its indexes, and the *Index Guide* (see p. 52), there are many pages explaining in detail policies and proper use, especially for the indexes. But the key to successful use of *CA*—as for any tool—is "hands-on" use on a regular basis. It requires both study and experience to use *CA* to full advantage.

CA has evolved to meet the changing needs of chemists and the significant increase in the volume and kinds of literature published. When *CA* editors began their drive toward computer-based production in the 1960s—and accelerated that drive in the early 1970s—the changes and benefits were numerous and significant, as are outlined later.

Clearly, a lengthy book could be written about *CA* alone, much less the balance of the important tools on chemical information. Accordingly, this chapter focuses on essentials, with emphasis on points less well understood by many chemists, on newer developments, and on important recent changes.

COVERAGE

The official statement of what *CA* covers is important and worth quoting in its entirety (6-3).

> It is the careful endeavor of *Chemical Abstracts* to publish adequate and accurate abstracts of all scientific and technical papers containing new information of chemical and chemical engineering interest and to report new chemical information revealed in the patent literature, but the American Chemical Society is not responsible for omissions or for such mistakes as may be made in abstracts and index entries.

Although the implication is that *CA* covers only new information, its editors cannot possibly check all material included to see if it is new. They can only make the "careful endeavor" noted in their policy statement. *CA* editors necessarily rely to a large extent on original author statements as to novelty of information in the literature.

ABSTRACT AND INDEX CONTENT

CA contains informative abstracts of the original documents. These abstracts are not, however, intended to replace the original, nor are they critical or evaluative reviews. Rather, the abstracts are filters, with the intent of providing the user with enough information on content to permit him to determine whether he wants to consult the originals. Any scientist who repeats a procedure based purely on the *CA* abstract is making a mistake.

Also, note that not all new substances and subjects reported in the original appear in the abstract; all are, however, covered by the volume and collective indexes, according to *CA* policy.

Recent (Volume 85) figures from CAS show that the overall subject and substance index "density" (number of index entries per abstract) is 6.5 for all 80 sections. Density for sections that pertain to synthetic chemistry—such as Section 28—is over 20 index entries. These figures are for general subject and chemical substance index entries only; they do not include additional access points such as author names and molecular formulas.

Note that *CA* has been significantly increasing index density since 1972; earlier years were not as deeply indexed.

During the ninth collective index period (and especially since Volume 82), indexing of reactants and intermediates has been extended to all subject areas covered by *CA*, thereby further strengthening the indexes. This is, of course, in addition to the products of reactions that are the usual focal points of interest.

SPEED OF COVERAGE AND INDEXING

The median time lag for the appearance of the abstract following issuance of the original publication is currently about 3.5 months; this includes completion of the indexing process which is important to on-line use (see Chapter 8). The current target for further improvement of this median is 3 months. Some chemists may believe that this is still too slow, but the abstracts may be available before the original journals are on library shelves, especially in the case of journals published abroad.

CA editors are making determined (and in many cases successful) efforts to further improve timeliness for all parts of *CA*. One notable example of success is speed of issuance of the semiannual (6-month) volume indexes. Speed of index appearance has improved dramatically to within about 6 months after the completion of the semiannual volumes. This is attributed in large measure to the massive computerization program initiated by *CA* editors in the 1960s.

CA INDEXES

The following is a broad summary of the contents of the volume and collective indexes to *CA* as currently published:

- ***Index Guide***—Details the major points of *CA* indexing policy for the appropriate collective period and provides cross-references from chemical substance names and general subject terms used in the literature to the terminology used. The *CA Index Guide* contains the following sections:

  Introduction, which describes cross-references, synonyms, and indexing policy notes listed in the main portion of the *Index Guide*.

  Volume Indexes to Chemical Abstracts: Organization and Use, which discusses the relationship between and use of the chemical substance, general subject, and formula indexes and index of ring systems.

  General subjects, which discusses the content of the general subject index.

  Selection of index names for chemical substances, which summarizes the rules used in deriving the *CA* index names for chemical substances.

Alphabetical sequence of cross-references, synonyms, and indexing policy notes, which is the main portion of the *Index Guide*.

Listing of general subject index headings in hierarchical order, which aids in generic searching.

- ***Chemical Substance Index***—relates the *CA* index names of chemical substances and their CAS registry numbers to *CA* abstract numbers for documents in which the substances are mentioned. Included with the *CA* index name is a brief description of the document's context.

- ***General Subject Index***—relates index entries which do not refer to specific chemical substances to the corresponding *CA* abstracts. These entries include concepts, general classes of chemical substances, applications, uses, properties, reactions, apparatus, processes, and biochemical and biological subjects.

- ***Formula Index***—relates the molecular formulas for chemical substances with the *CA* chemical substance index names, CAS registry numbers, and corresponding *CA* abstract numbers.

- ***Index of Ring Systems***—lists ring composition, ring size, and number of rings thus providing a means for determining the systematic *CA* index names for specific ring systems as well as the nonsystematic *CA* index names for cyclic natural products containing these ring systems.

- ***Author Index***—lists authors, patentees, and patent assignees in alphabetical order with titles of their articles or patent specifications and *CA* abstract numbers.

- ***Numerical Patent Index***—relates new patent numbers, grouped by country in ascending numerical order, with their corresponding *CA* abstract numbers.

- ***Patent Concordance***—links equivalent patents (see p. 151) to one another and to the abstract number of the first patent received and abstracted in *CA*.

Figure 9 depicts the development of *CA* Collective Indexes. Each weekly issue of *CA* also contains several indexes. The issue indexes include the following:

- 1st through 4th decennial indexes each contain author and subject indexes.
- 1st collective formula index covers 1920–1946.
- Ten-year numerical patent index covers 1937–1946.
- 5th collective index contains author, subject, numerical patent, and formula indexes.
- 6th collective index contains author, subject, numerical patent, and formula indexes and index of ring systems.
- 7th collective index contains author, subject, numerical patent, and formula indexes; index of ring systems; and patent concordance.
- 8th collective index contains author, subject, numerical patent, and formula indexes; index of ring systems; patent concordance; and index guide.
- 9th collective index contains author, chemical substance, general subject, numerical patent, and formula indexes; index of ring systems; patent concordance; and index guide.

Figure 9 Development of CA collective indexes.

a Keyword index
b Numerical patent index
c Patent concordance
d Author index

The chemist can use the keyword index to search specific issues of *CA* for subject content. The vocabulary is uncontrolled, that is, nonsystematic and not standardized. This permits the index to be produced quickly and published as a part of each issue, but it makes for substantially less reliability than is found in the semiannual volume indexes, which appear some 6 months after completion of the volume. The keyword index is an important and useful tool in on-line searching (see p. 89).

The numerical patent index, patent concordance, and author index are cumulated semiannually in the volume indexes and every 5 years in the collective indexes.

The section on nomenclature, which follows, contains further information pertinent to *CA* index use.

CA NOMENCLATURE

Beginning with Vol. 76 (1972) CAS made a significant change in index nomenclature policy. *CA* now uses the more fully systematic IUPAC (International Union of Pure and Applied Chemistry) nomenclature in almost all cases for main chemical substance headings. "Trivial" or "nonsystematic" names are used only rarely. Note that this change applies to the bold face headings, not to heading modifications in which original author nomenclature is used. Some examples follow:

Old Policy	***New Policy***
acetic acid	no change
aniline	benzenamine
carbonic acid	no change
ethylene	ethene
ethylenediamine	1,2-ethanediamine
ethylenimine	aziridine
ethylene glycol	1,2-ethanediol
ethylene oxide	oxirane
ethyl ether	ethane, 1,1′-oxybis-
ethyl methyl ketone	2-butanone
formic acid	no change
phenol	no change

Some chemists believe that this policy change makes it more difficult for them to use *CA*. They argue that such heavily studied chemicals as aniline, for example, should be indexed under that name and not under benzenamine which few chemists use in conversation, normal reporting, or commerce. Although the fully systematic name for aniline is cross-referenced from aniline in the *CA Chemical Substance Index*, most trivial names are not similarly cross-referenced, except in the *Index Guide*.

CA could not afford to reproduce *all* cross-references in every volume index—hence the *Index Guide* was born. In recent volumes, *CA* editors have reinserted some 400 to 500 cross-references into the index as guides for some of the most referenced substances. This was done in response to user requests.

CA editors believe that their index name policy change is fully justified.

The primary intent of this effort is to group structural families together in the alphabetically ordered *CA Volume Chemical Substance Index*. Other reasons behind the use of more fully systematic index names include:

1 They are generated more consistently.

2 They are translated more readily to complete structures.

3 They are edited and verified more rapidly and consistently.

4 They reduce manual and machine search efforts.

5 They employ fewer and more systematic nomenclature principles.

A simple example of past confusion is 4-methoxytoluene, also known as 4-methylanisole. Anisole and toluene are both descriptive "trivial" names. Under the new fully systematic index policies, there is no confusion. The compound is now designated by *CA* as 1-methoxy-4-methylbenzene.

CA editors have provided several powerful tools to help chemists adjust to the new nomenclature and make most effective use of the general subject and chemical substance indexes published under the new policies. These include:

a The Index Guide (IG)

b The Formula Index

c The Chemical Registry System and Registry Handbook

As previously indicated, the IG, to which supplements are published annually, is a collection of cross-references, explanations of headings, and synonyms, which should be consulted if the current *CA* index policy for a substance is not known. (Prior to the initiation of the IG, the chemist found such cross-references and notes scattered throughout the *Subject Index*). The IG is now the indispensable key to efficient use of the *General Subject* and *Chemical Substance Indexes*, with special value in helping the chemist "translate" trivial names to the more fully systematic index names used by *CA*. Chemists using *CA* should use the most recent IG supplement first and then the IG for the ninth collective period (1972–1976). There is also an IG for the eighth collective period (1967–1971), but much of the latter IG does not apply to current practice.

The *Formula Index*, published semiannually and cumulated every 5 years, provides the chemist with the quickest and most reliable route of access to names of individual substances as listed in the *Chemical Substance Index*. The *Formula Index* is easier and more straightforward to use than the IG, especially if an accurate molecular formula is already in hand. Under each molecular formula, the chemist will find the *CA* systematic name for the compound, the registry number (for positive identification), and an abstract number in which the compound is mentioned in *CA* issues. A physical property, such as boiling point, is given when an original document does not provide enough information for an unambiguous name. Chemists should, however, not let the above comments discourage them from use of the IG; calculation of molecular formulas can also be error-prone.

The next step after use of the *Formula Index* is to consult the *Chemical Substance Index* if all entries for a given substance are needed. The *Chemical Substance Index* also provides references for derivatives of the substance in question.

The CAS Chemical Registry System, developed by CAS beginning in 1965, reached some 4 million entries during 1977. In this system each substance is assigned a unique, unambiguous computer record that is independent of the vagaries of nomenclature. For example, the Registry Number for benzene is 71-43-2.

CAS registry numbers have been widely accepted by the chemical publishing community and are used in many journals, in such tools as *HEEP* (see p. 176), *TOXLINE* (see p. 166), and in many other places.

With a registry number in hand, the chemist can consult *CA*'s *Registry Handbook*, and its supplements, to locate the *CA* index name and the molecular formula.

A few registry numbers may change, particularly when the original document (article, patent, etc.) supplied incomplete or partial information. In some cases multiple registry numbers may be assigned to what later proves to be the same structure. As CAS learns more about such structures, old numbers are replaced by the preferred numbers. These are reported in the supplements.

Note that it is not possible to find associated *CA* abstracts for all CAS registry numbers, because some of these numbers resulted from the registration of special data collections such as *The Colour Index*.

PARENT COMPOUND HANDBOOK

Another new tool available from CAS is the *Parent Compound Handbook*. This tool, which first appeared in 1977, updates and expands the coverage of *The Ring Index* (and its supplements) which it replaces. Uses include the following:

- The *Parent Compound Handbook* contains information about all ring systems and natural products presently in the CAS *Chemical Registry System*.
- Supplements to the *Parent Compound Handbook* provide a current awareness service for alerting to new ring systems and natural products entered into the CAS Chemical Registry System during processing of the primary literature for *CA*.

▪ It helps determine the *CA* index name of a ring system or natural product; the *CA* index name can then be used to enter the *Chemical Substance Index* to *CA* to find references for documents reporting substances containing the ring system or natural product.

▪ It provides a bridge between CAS systematic nomenclature and Wiswesser line notations (see p. 74).

▪ When searching computer-readable files, the chemist can use information from the *Parent Compound Handbook* to formulate substructure search strategies—that is, strategies which allow one to search for a group of substances that have in common certain ring systems or products of interest.

The *Parent Compound Handbook* includes for a parent compound:

1 Chemical structural diagram illustrating the nomenclature locant numbering system.

2 CAS registry number.

3 Current *CA* index name used in *Chemical Substance Index* to *CA*.

4 Molecular formula.

5 Wiswesser line notation (for ring parents, cyclic stereoparents, and some acyclic stereoparents).

6 Ring analysis data: number of rings, size of rings, and elemental analysis of rings (for ring parents and cyclic stereoparents).

7 *CA* reference (if the parent compound was referenced in a *CA* abstract after Volume 78 (1973).

REGISTRY HANDBOOK—COMMON NAMES

Beginning in 1977, CAS initiated the *Registry Handbook—Common Names* on microform as a new experimental service. Whether the service will become permanent is not known at this writing. This new tool provides access to CAS registry numbers through the variety of substance names that are commonly used in the chemical literature and in chemical and allied industries. Molecular formulas are also given when known.

There are two parts: the name section and the number section. Emphasis is on simple, less systematic names, such as so-called author names, literature names, common names, and trivial names. The chemist can use the number section to identify sets of synonymous common names for the same substance.

USE OF SUBDIVISIONS

For hundreds of commonly reported chemicals and some classes of chemicals—such as alkanes, benzene, boric acid, propane, 1-propanol, pyridine, steel, sugars, sulfuric acid and tellurium—for which there are many index references, the chemist will find his searching task simplified by subdivision of the index heading into these general categories (called "qualifiers") in *CA*:

Analysis
Biological studies
Occurrence
Preparation
Properties
Reactions
Uses and miscellaneous

This practice was initiated with the *Eighth Collective Index* period (1967–1971).

For years prior to 1967, and for compounds for which the index is not subdivided as shown above, searching the indexes can become both onerous and time-consuming. The chemist or engineer, if he wishes to ensure absolute completeness of his search, should read all index entries under headings that are not subdivided. This advice holds true for important searches, but does not necessarily apply to "everyday" searches.

Even subsequent to use of subdivisions of the heading, the chemist can never be completely sure that a search under any one of the subdivisions will yield every item of desired information. A document that emphasizes one aspect of a subject may also have valuable data on other aspects. For example, in the indexing of a document on "preparation" and "properties" of a substance, either of these subdivisions may be used. The *CA* document analyst uses his judgment as to which subdivision best fits emphasis of the document. Accordingly, the user may need to consult both subdivisions to retrieve all of what is needed. Note that the subdivision "preparation" spans the range from laboratory synthesis through commercial-scale manufacturing processes.

The subdivision system just described can be a valuable time saver. Hopefully, the system can eventually be extended to include virtually all chemicals indexed by *CA*.

In addition to the subdivisions described above, index headings with large numbers of entries are also divided into chemical functional group subdivisions. These include:

Acetals
Anhydrides
Anhydrosulfides
Compounds
Derivatives (general)
Esters
Ethers
Hydrazides
Hydrazones
Lactones

Mercaptals
Mercaptoles
Oxides
Oximes
Polymers

SIZE

The sheer physical bulk of *CA* is imposing, almost forbidding to some. The indexes, in particular, are so large that considerable time can be required to locate what is needed. For example, over 80 books will make up the *Tenth Collective Index*. A single issue of *CA* can total some 700 pages. But if *CA* is to maintain its objective of covering virtually all new chemical information, the current massive issues and indexes need to be maintained. Use of *CA* on microfilm and of the computer-readable versions (see p. 82) can help cope with this size/bulk situation. Familiarity with *CA*, and experience in its use greatly facilitates speed and ease of use.

Figure 10 summarizes some trends in the growth of *CA*. Preliminary indications are that the relative rate of growth of chemical literature, as covered by *CA*, may be slowing down somewhat.

COMPLEXITY

The development of the science of chemistry, coupled with the use of large numbers of automatic laboratory devices, have combined to produce substantially more data in recent years. This requires additional routes of access by CAS if the total literature is to be covered. *CA* today contains more data and more indexing results than ever before. The tools now

Collective Index	Years Covered	CA Volumes	Number of Documents Referenced
1st	1907–1916	1–10	192,000
2nd	1917–1926	11–20	218,000
3rd	1927–1936	21–30	544,000
4th	1937–1946	31–40	513,000
5th	1947–1956	41–50	663,000
6th	1957–1961	51–55	637,000
7th	1962–1966	56–65	1,024,000
8th	1967–1971	66–75	1,466,000
9th	1972–1976	76–85	2,024,000
10th	1977–1981	86–95	(2,600,000)

Figure 10 ***CA Index growth. (Number of documents estimated for 1977–1981.)***

published by *CA* are, of themselves, not more complex. It is simply a matter of more tools the user needs to master that, in the long run, may make the task easier.

PATENT COVERAGE

CA does not cover all new chemical patents in its abstracts. For example, if a patent is published first in a country other than the United States, and subsequently issued in the U.S. (or any other country), these "equivalents" (see p. 151) will not be reabstracted. Rather, they will appear only in the patent concordance to *CA*. This policy, initiated in 1962, saves time and funds for *CA*, and also saves additional price increases for subscribers, but users need to spend considerably more time and effort to locate equivalent patents. See also Chapter 11 on patents.

Note that, overall, *CA* patent coverage has improved significantly, especially since about 1960. Prior to that, patent coverage is spotty, and, at least for composition of matter patents, *Chemisches Zentralblatt* is a better source.

REVIEWS

CA provides excellent coverage of review papers. Their definition of a review is based primarily on whether the author of the original document so designates it. If new experimental information is also provided, however, the document is not treated as a review, but as a regular research report and is indexed thoroughly.

CHEMICAL MARKETING AND BUSINESS INFORMATION

This is not usually included directly in *CA*. It is "relegated" to the subsidiary publication issued by Chemical Abstracts Service (CAS) known as *Chemical Industry Notes (CIN)* (see also p. 220). CIN is excellent, but its abstracts (better termed extracts) and indexes are not as good as the full *CA*. Material appears in CIN quickly—within just a few weeks after the original. A typical issue contains the following sections:

Production
Pricing
Sales
Facilities
Products and processes
Corporate activities
Government activities
People
Keyword index
Corporate name index

DISSERTATIONS

Some chemists believe that an example of an area for improvement is that of dissertations. Rather than provide an abstract of the dissertation, the user is usually referred by *CA* to *Dissertation Abstracts* (see p. 36). This requires an often inconvenient second step, since *Dissertation Abstracts* is not readily available in many chemistry libraries and information centers. But there is no other way. *CA* could never assemble copies of all dissertations for their own abstracting—and particularly, indexing. The content of *Dissertation Abstracts* is too long for *CA* use, and *CA* editors are reluctant to modify abstracts without having the originals in front of them.

COVERAGE OF DOCUMENTS FROM THE SOVIET UNION

Many Soviet documents are fully abstracted and indexed by *CA*. Some, however, are accessible only through the abstracting and indexing service *Referativnyi Zhurnal (Ref. Zh.)* Because the abstracts in *Ref. Zh.* are copyrighted, only the titles and bibliographic information (including the *Ref. Zh.* data) are published in *CA*. The statement "Title only translated" is used instead of an abstract. Keywording and indexing are done from the title only.

AVAILABILITY OF MATERIALS COVERED BY CA

This is a serious problem which is being studied by various groups as this book is written. It is estimated that about 10% of material covered by *CA* is not readily available to the North American chemist at present. A variety of solutions have been proposed by the different groups. An example is establish-

BIOCHEMISTRY SECTIONS

1. Pharmacodynamics
2. Hormone Pharmacology
3. Biochemical Interactions
4. Toxicology
5. Agrochemicals
6. General Biochemistry
7. Enzymes
8. Radiation Biochemistry
9. Biochemical Methods
10. Microbial Biochemistry
11. Plant Biochemistry
12. Nonmammalian Biochemistry
13. Mammalian Biochemistry
14. Mammalian Pathological Biochemistry
15. Immunochemistry
16. Fermentations
17. Foods
18. Animal Nutrition
19. Fertilizers, Soils, and Plant Nutrition
20. History, Education, and Documentation

ORGANIC CHEMISTRY SECTIONS

21. General Organic Chemistry
22. Physical Organic Chemistry
23. Aliphatic Compounds
24. Alicyclic Compounds
25. Noncondensed Aromatic Compounds
26. Condensed Aromatic Compounds
27. Heterocyclic Compounds (One Hetero Atom)
28. Heterocyclic Compounds (More Than One Hetero Atom)
29. Organometallic and Organometalloidal Compounds
30. Terpenoids
31. Alkaloids
32. Steroids
33. Carbohydrates
34. Synthesis of Amino Acids, Peptides, and Proteins

MACROMOLECULAR CHEMISTRY SECTIONS

35. Synthetic High Polymers
36. Plastics Manufacture and Processing
37. Plastics Fabrication and Uses
38. Elastomers, Including Natural Rubber
39. Textiles
40. Dyes, Fluorescent Whitening Agents, and Photosensitizers
41. Leather and Related Materials
42. Coatings, Inks, and Related Products
43. Cellulose, Lignin, Paper and Other Wood Products
44. Industrial Carbohydrates
45. Fats and Waxes
46. Surface—Active Agents and Detergents

APPLIED CHEMISTRY AND CHEMICAL ENGINEERING SECTIONS

47. Apparatus and Plant Equipment
48. Unit Operations and Processes
49. Industrial Inorganic Chemicals
50. Propellants and Explosives
51. Fossil Fuels, Derivatives and Related Products
52. Electrochemical, Radiational, and Thermal Energy Technology
53. Mineralogical and Geological Chemistry
54. Extractive Metallurgy
55. Ferrous Metals and Alloys
56. Nonferrous Metals and Alloys
57. Ceramics
58. Cement and Concrete Products
59. Air Pollution and Industrial Hygiene
60. Sewage and Wastes
61. Water
62. Essential Oils and Cosmetics
63. Pharmaceuticals
64. Pharmaceutical Analysis

PHYSICAL AND ANALYTICAL CHEMISTRY SECTIONS

65. General Physical Chemistry
66. Surface Chemistry and Colloids
67. Catalysis and Reaction Kinetics
68. Phase Equilibriums, Chemical Equilibriums, and Solutions
69. Thermodynamics, Thermochemistry, and Thermal Properties
70. Nuclear Phenomena
71. Nuclear Technology
72. Electrochemistry
73. Spectra by Absorption, Emission, Reflection, or Magnetic Resonance, and Other Optical Properties
74. Radiation Chemistry, Photochemistry, and Photographic Processes
75. Crystallization and Crystal Structure
76. Electric Phenomena
77. Magnetic Phenomena
78. Inorganic Chemicals and Reactions
79. Inorganic Analytical Chemistry
80. Organic Analytical Chemistry

Guidelines used to assign abstracts to CA Sections according to their subject content are summarized in the publication, SUBJECT COVERAGE AND ARRANGEMENT OF ABSTRACTS BY SECTIONS IN CHEMICAL ABSTRACTS (Subject Coverage Manual).

Figure 11 **CA *Section Groupings***

ment of one or more centralized facilities where availability of most or all materials covered by *CA* will be "guaranteed"; this is being considered, as are other options.

A cooperative arrangement with the U.S. National Technical Information Service is expected to help make available copies of articles from the most frequently cited journals. See also Chapter 5.

MISTAKES

Mistakes, though infrequent, will be found in the abstracts, indexes, and other publications of *CA*. *CA* staff members are human, and perfection is too much to ask for in such a large and complex work. Also, huge volumes of literature must be handled in relatively short periods of time. Correspondence with the editors of *CA* can almost always clear up any suspected mistake or other discrepancy.

COST

At a $4200 subscription fee per year (1978), the full *CA* is now priced far beyond the pocketbooks of almost all chemists as individuals. It has become an "institutional" publication which only organizations with good budgets can afford. But certain of its parts, for example, the individual *Section Groupings* (see Figure 11) are available at prices as low as $35 per year for ACS members.

CA officials note, and correctly so, that the current price of the full *CA* is necessary if it is to maintain its high standards and broad coverage and remain a viable service to the chemical community. Even so, *CA* is a bargain because of its broad scope and high quality.

FUTURE

The most likely outlook for *CA* is further emphasis on use of computers in its production and especially in its use. Another potential trend is the initiation of additional new publications, such as *CA Selects* to meet special needs. Should *CA* get much bigger, more complex, and significantly more expensive, we can look for further proliferation of smaller, specialized services targeted to meet the needs of groups with specific interests (see Chapter 7).

7 OTHER ABSTRACTING AND INDEXING SERVICES

Although *CA* is, and will remain, the bulwark of the chemist's information armanentarium, other abstracting and indexing services of different scope or content are useful. They offer the chemist unique or supplementary benefits as compared to *CA* in certain situations—benefits which the chemist can often use to good advantage.

Let us first consider several examples published in the United States. One is found in the work of the Institute of Paper Chemistry, Appleton, WI 54911. The Division of Information Services of the institute issues the following publications:

Abstract Bulletin (available in printed, microform and on-line versions)
Keyword Index to the Abstract Bulletin
Bibliographic Series
Conference Proceedings

The division also offers the following types of services:

Computer tapes and software
Current awareness searches
Custom literature searches
Dissertations (loans of copies)
Library, including photocopy service
Reprints

Retrospective computer searches
Translations

From the above listing, it should be clear that the institute offers a complete array of services.

For the chemist doing research in paper chemistry, the Institute's *Abstract Bulletin* can have a number of advantages. For example, the abstracts can be slanted specifically toward interests of chemists working in this field. The index can be easier to use than a larger index because its scope is limited to paper chemistry, and because specifics of interest to the paper chemist can be indexed in detail. This can mean quicker and easier access to needed information. Also, some highly specialized materials that could not be justified in a larger service of broad scope are more likely to be found in a publication such as this. Another potential advantage is the relative ease with which publications like this can be scanned, as compared to some of the more bulky sources.

As another case, consider *Fertilizer Abstracts* published by the Tennessee Valley Authority, Muscle Shoals, AL 35660. This abstracting and indexing service covers the chemistry and technology of plant foods and the chemicals that go into their manufacture. Accordingly, chemists interested primarily in such products as phosphoric acid and ammonia may find *Fertilizer Abstracts* the most satisfactory indexing and abstracting source. The indexes are detailed, comprise a highly manageable number of pages, and are easy to use. The abstracts are collected together in a compact monthly booklet. All of the material is easily scanned for either current awareness or retrospective use.

There are a number of other "smaller" services, especially in the more applied aspects of chemistry. These are too numerous to mention here, but the two cited above are sufficient to illustrate potential advantages. Several important relatively large services which deserve the attention of the chemist are noted in the paragraphs which follow.

INSTITUTE FOR SCIENTIFIC INFORMATION

The Institute for Scientific Information (ISI) is an outstanding for-profit organization which offers a wide variety of services to the chemist. ISI is located at 325 Chestnut Street, Philadelphia, PA 19106. The success of this organization is due in large measure to the innovativeness of its founder and chief executive, Dr. Eugene Garfield. Some of its publications and other services are listed throughout this book. Several related to abstracting and indexing are described in the following paragraphs.

Current Abstracts of Chemistry (CAC) is a highly regarded weekly abstracting service published by the Institute for Scientific Information (ISI) which tells what *new* organic compounds and syntheses are reported in about 110 of the "most important" journals to organic chemists. Over 170,000 new compounds (including intermediates) will be reported in 1977; over two million compounds have been recorded since 1960. ISI claims that this is nearly 90% of all new organic compounds reported in the journal literature. Coverage is prompt; ISI says that most articles appear in *CAC* within 45 days after the article is published. *CAC* provides the following information:

1 Abstract number for easy look-up in the indexes.

2 Complete citation of the original reference.

3 Author's abstract from the original article.

4 Compound applications, that is, testing of new compounds for specific uses.

5 ISI accession number, which permits ordering copy of the article through ISI's OATS service.

6 Information at a glance on what analytical techniques were used.

7 Extensive use of structural and reaction flow diagrams; this is one of the most significant features of *CAC*.

8 Author's number for all compounds to speed examination of abstract and original article.

Index Chemicus is the weekly index to *CAC*. It is cumulated quarterly and annually. Key access points in this index include subject, molecular formula, organization, and author. Other indexes include journal, biological activities, instrumental data, and alert of new reactions, new syntheses, and labeled compounds.

Chemical Substructure Index permits substructure searching of compounds reported in *CAC*. Access is provided through use of the so-called Wiswesser line notation (WLN). WLN is an alphanumeric linear encoded representation of structures. It is a system that is probably most fully used by ISI, but it is also used by many other organizations and in numerous publications. William J. Wiswesser developed the system which bears his name some 25 years ago; improvements have been introduced on an ongoing basis ever since. Some advantages claimed for WLN include: independence from the vagaries of some nomenclature systems; relatively easy search for structures and substructures, and use either manually or via computer. Decoding (reading) of WLN is said to be learned relatively easily. Encoding of structures is claimed to be possible after about a month of study and practice.

WLN is not always as unambiguous as some would like, and the system is still evolving to handle such problem areas as inorganics. An excellent manual (7-1) is available for those interested in further details, and ISI offers other useful instructional aids.

The strong points of *CAC* include speed of coverage and easy-to-read, highly graphic abstracts which include structural and reaction flow diagrams. The indexes are another unique feature.

The chemist should note, however, that *CAC* is limited to journals; patents are presently excluded. This policy excludes compounds and syntheses reported in patents only.

Also, as noted, although a large number of important journals are covered, many are not. However, studies at ISI indicate that most new compounds are reported in only a relative handful of journals. One further point of interest is that much new information, such as new uses about *old* compounds is excluded from *CAC*; the reason for this is that the focal point of *CAC* is primarily on *new* compounds, and new reactions and syntheses.

A service associated with *CAC* is *ANSA*, *Automatic New Structure Alert*. This is a monthly computerized alerting service. It provides current awareness and retrospective search capability for new organic compounds reported in *CAC* that contain specified chemical substructures. This file goes as far back as 1966; plans call for extension back to 1960. The output is a computer print giving the following information for each article in which a compound of interest is reported:

1 Full title
2 Author
3 Journal citation
4 Author address
5 Analytical techniques used
6 Subject indexing terms applicable to this article
7 Abstract number referring back to *CAC*

It is not necessary to be a subscriber to *CAC* to use *ANSA*.

Also available is the *Index Chemicus Registry System* (ICRS). This monthly magnetic tape service permits an organization to incorporate into its own computer system the *CAC* and *Index Chemicus* data base.

Another ISI publication is the multidisciplinary *Science Citation Index* (SCI). *SCI* is available in printed form and also in an on-line computer-based version known as *SCISEARCH* through Lockheed Information Systems (see Chapter 8).

SCI indexes material covering over 2600 journals representing more than 100 disciplines, including chemistry. Indexing is via:

a ***Citation Index***—This important tool lists documents referred to (cited) in the current literature alphabetically by name of author. It permits the user who already knows one or more authors who have written in a field of interest to identify other items citing the previously known pertinent work. This can lead to a network or chain of related references. For example, a user may wish to know who has cited his own work or the work of colleagues.

b ***Source Index***-This index is alphabetically arranged by author, and there is also a separate index by organization. It is a way of keeping track of work by specific authors and organizations.

c ***Permuterm Subject Index***-This is an index based on the original words in the titles of items covered. All significant words in a title are coupled or paired together as shown in Figure 12.

SCI is published three times a year and cumulated annually. Cumulations go back to 1961. There are also 5-year cumulations covering 1965–1969 and 1970–1974. It can be a unique and powerful tool when properly used. The most significant feature is the citation indexing employed.

SERVICES PUBLISHED OUTSIDE THE UNITED STATES

A number of important abstracting and indexing services of interest to the chemist are, or have been , published outside the United States.

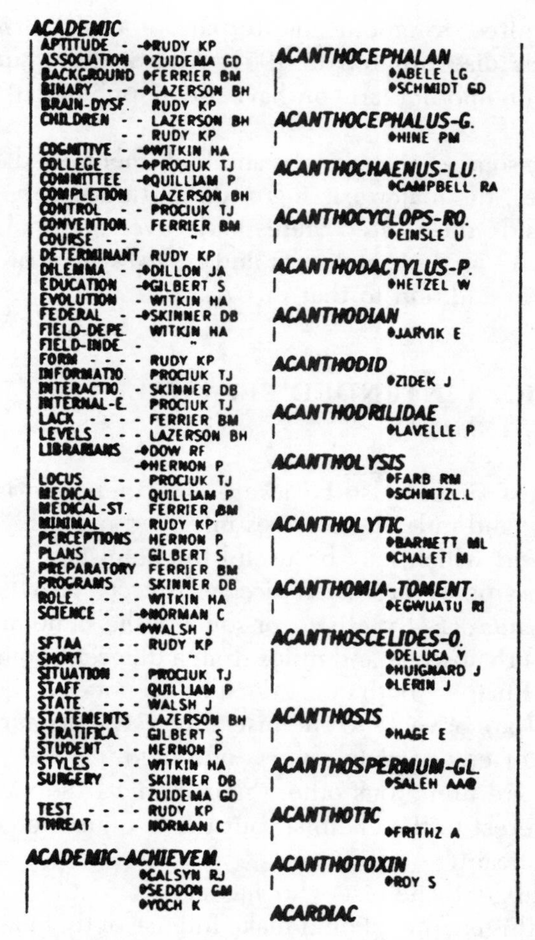

ACADEMIC
APTITUDE - -•RUDY KP
ASSOCIATION •ZUIDEMA GD
BACKGROUND •FERRIER BM
BINARY - - -•LAZERSON BH
BRAIN-DYSF. RUDY KP
CHILDREN - - LAZERSON BH
- - RUDY KP
COGNITIVE - -•WITKIN HA
COLLEGE - -•PROCIUK TJ
COMMITTEE -•QUILLIAM P
COMPLETION LAZERSON BH
CONTROL - - PROCIUK TJ
CONVENTION. FERRIER BM
COURSE - - - "
DETERMINANT RUDY KP
DILEMMA - -•DILLON JA
EDUCATION -•GILBERT S
EVOLUTION - WITKIN HA
FEDERAL - -•SKINNER DB
FIELD-DEPE. WITKIN HA
FIELD-INDE. - "
FORM - - - - RUDY KP
INFORMATIO. PROCIUK TJ
INTERACTIO. - SKINNER DB
INTERNAL-E. PROCIUK TJ
LACK - - - - FERRIER BM
LEVELS - - - LAZERSON BH
LIBRARIANS -•DOW RF
-•HERNON P
LOCUS - - - PROCIUK TJ
MEDICAL - - QUILLIAM P
MEDICAL-ST. FERRIER BM
MINIMAL - - RUDY KP
PERCEPTIONS HERNON P
PLANS - - - GILBERT S
PREPARATORY FERRIER BM
PROGRAMS - SKINNER DB
ROLE - - - - WITKIN HA
SCIENCE - - -•NORMAN C
- - -•WALSH J
SFTAA - - - RUDY KP
SHORT - - - "
SITUATION - - PROCIUK TJ
STAFF - - - QUILLIAM P
STATE - - - WALSH J
STATEMENTS LAZERSON BH
STRATIFICA. - GILBERT S
STUDENT - - HERNON P
STYLES - - - WITKIN HA
SURGERY - - SKINNER DB
- - ZUIDEMA GD
TEST - - - - RUDY KP
THREAT - - - NORMAN C

ACADEMIC-ACHIEVEM.
•CALSYN RJ
•SEDDON GM
•YOCH K

ACANTHOCEPHALAN
•ABELE LG
•SCHMIDT GD

ACANTHOCEPHALUS-G.
•HINE PM

ACANTHOCHAENUS-LU.
•CAMPBELL RA

ACANTHOCYCLOPS-RO.
•EINSLE U

ACANTHODACTYLUS-P.
•HETZEL W

ACANTHODIAN
•JARVIK E

ACANTHODID
•ZIDEK J

ACANTHODRILIDAE
•LAVELLE P

ACANTHOLYSIS
•FARB RM
•SCHNITZL.L

ACANTHOLYTIC
•BARNETT ML
•CHALET M

ACANTHOMIA-TOMENT.
•EGWUATU RI

ACANTHOSCELIDES-O.
•DELUCA Y
•HUIGNARD J
•LERIN J

ACANTHOSIS
•HAGE E

ACANTHOSPERMUM-GL.
•SALEH AAQ

ACANTHOTIC
•FRITHZ A

ACANTHOTOXIN
•ROY S

ACARDIAC

Figure 12 ***Excerpt from Permuterm Index to*** **Science Citation Index.**

The ongoing Russian service (*Ref. Zh.*) was referred to in the previous Chapter.

Also mentioned in Chapter 6 was *Chemisches Zentralblatt (C.Z.)* which was discontinued as such in 1969. *C.Z.* is the "grandfather" of chemical abstracting services, with origins traceable back to 1830.

The United Kingdom counterpart of *C.Z.*, *British Abstracts*, was discontinued in 1953. Additionally, services of one kind or another are or have been published in other countries.

As with some of the services and tools mentioned earlier in this chapter, the major existing or discontinued services published outside the United States may cover material not covered by *CA*, or their abstracts and indexes may provide information in addition to that in *CA*.

SERVICES INTENDED PRIMARILY FOR NONCHEMISTS

The chemist should also be aware that there are important abstracting and indexing services intended primarily for nonchemists *but* which may be useful to chemists.

Two examples are *Biological Abstracts (BA)* and *Engineering Index (EI)*. Both cover some of the same material as *CA*. But both abstract and index from a different point of view than *CA*. Further, both cover documents not covered by *CA*. *BA* and *EI* are of value to chemists interested primarily in the biological or engineering aspects of a field.

There are numerous other "nonchemical" services of potential interest to the chemist, but the two just noted are the most significant.

The diligent chemist who has access to the tools mentioned in this chapter should make full use of them as needed.

8 COMPUTER-BASED ON-LINE AND OFF-LINE INFORMATION RETRIEVAL SYSTEMS AND SERVICES

An on-line information retrieval system is one in which a user can, via computer, directly interrogate a machine-readable data base such as the indexes to *CA*.

There is two-way communication between the user and computer by way of input/output devices such as a teletypewriter or cathode ray tube display connected to the computer by a channel which may be a telephone line.

WHY USE ON-LINE SERVICE

On-line information retrieval offers the chemist totally new dimensions and unique advantages in searching of patent, journal, and report literature. Rapid, fingertip access to several million references from a wide variety of sources (primarily abstracting and indexing services) is made possible at highly competitive costs, as compared to other methods of searching. The systems available today often permit users to identify recent pertinent references more quickly and effectively than is possible by using more conventional manual techniques.

The concept, as applied to chemistry, is relatively new. It did not win widespread acceptance in the chemical community until about 1973.

The purpose of this chapter is to give the modern chemist some knowledge of how on-line systems function. It will also suggest strategic concepts that should be considered in searching of traditional (printed) tools (see also Chapter 3).

SOME CHARACTERISTICS OF ON-LINE SEARCHING

On-line searching is based on use of a local interactive computer terminal connected to computer centers by telephone either directly or via special network lines. The user can ask a question, and a few seconds later the computer will respond, almost as in a conversation.

Fortunately for the chemist, almost all major information services of chemical interest are available on-line through several suppliers, most of whom provide access throughout the United States and in other countries. Examples of suppliers or developers are:

Bibliographic Retrieval Services, Inc. (BRS)
1462 Erie Boulevard
Schenectady, NY 12305

Fein-Marquart Associates, Inc.
7215 York Road
Baltimore, MD 21212
(Chemical Information System developed jointly with National Institutes of Health and Environmental Protection Agency.)

Lockheed Information Systems
3251 Hanover Street
Palo Alto, CA 94304

National Library of Medicine
8600 Rockville Pike
Bethesda, MD 20014

SDC Search Service (System Development Corporation)
2500 Colorado Avenue
Santa Monica, CA 90406

One example of material available on-line is the indexes to the weekly issues of *CA*. Scanning these valuable indexes manually over a period of weeks or months may be almost prohibitively time consuming. But use of on-line systems permits rapid scanning and also offers other important benefits.

It enables the user to create and use, as necessary, a search logic that is more complex than can be handled manually. This is done by combining key terms (such as subjects) used in searching, as shown in the following examples:

1 *A* or *B* or *C* or *D*
2 *A* & *B* & *C* & *D*
3 (*A* & *B*) or (*C* & *D*)

Similarly, the user can specify that only certain types of publications are wanted, such as patents issued in a specific country or language within a given time. All other materials could be automatically excluded in an on-line search. Doing this kind of specification in a conventional manual search is possible, but much more time consuming.

On-line systems also make it possible for the chemist to search a wide variety of sources and tools which might otherwise not be available because of geographical, budgetary, or other limitations.

SOURCES OF INTEREST TO CHEMISTS THAT ARE SEARCHABLE ON-LINE

The partial list of sources in Figure 13 is representative as of 1977. New sources (or data bases as they are called) are added and others are dropped periodically, depending on availability, demand, and interest or need.

The chemist should consult his local chemical information specialist or chemistry librarian for details on what is currently available. A complete list might number more than 50 data bases.

Note that, as mentioned elsewhere in this chapter, in only a few instances is the complete data base available for search on-line. Nevertheless, so much of importance is on-line that

1. *Biological Abstracts*
2. *Chemical Abstracts*
3. *Chemical Industry Notes*
4. *Derwent World Patents Index and Central Patents Index*
5. *Engineering Index*
6. *IFI Index to U.S. Chemical Patents (Claims ™ Chem)*
7. *Science Citation Index ® (SCISEARCH ®)*
8. Smithsonian Science Information Exchange

Figure 13 Some tools of chemical interest searchable on-line.

the significance of this mode of search cannot be overemphasized.

ON-LINE OUTPUT

On-line output usually consists of a list of references, including full bibliographic citations and indexing terms. In some cases, abstracts are also available as on-line output. In one or two cases, copies of original documents can be ordered while the user is on-line. See Figure 14 for examples of input and output. The future outlook is for more "hard" data, not just references, to be available on-line.

USE OF ON-LINE SERVICES

The on-line system concept is not a panacea. Most data bases are available on-line only since about 1970–1972, although some go back much earlier. This means that, for complete coverage—going back to the "beginning"—the chemist also

```
PROG:

SS 1 /C?
USER:
COAL AND ALL LIQUE:

PROG:
SS 1 PSTG (347)

SS 2 /C?
USER:
COAL AND ALL GASIF:

PROG:
SS 2 PSTG (1076)

SS 3 /C?
USER:
1 OR 2

PROG:
SS 3 PSTG (1335)

SS 4 /C?
USER:
3 AND 7726 (UP)

PROG:
SS 4 PSTG (32)

SS 5 /C?
USER:
4 AND NOT P (DT)

PROG:
TIME OVFLW: CONT? (Y/N)

USER:
Y

PROG:
SS 5 PSTG (18)

SS 6 /C?
USER:
"PRINT 18

PROG:

-1-
AN  - CA087262057490
TI  - CLEANING-UP THE ENVIRONMENT BY MINIMIZING SULFUR DIOXIDE
      EMISSIONS
AU  - JILEK, JAROMIR
OS  - GASVERBUND OSTSCHWEIZ A.-G., ZURICH
SO  - GAS, WASSER, ABWASSER 57 6 457-64
PY  - 77
LA  - GER
CC  - CA059002
```

(Figure 14—to page 86)

```
-2-
AN  - CA08726204099V
TI  - MAKING LIQUIDS FROM SOLID FUELS
AU  - TAUPITZ, K.C.
OS  - IND. DEV. CORP., SALISBURY
SO  - HYDROCARBON PROCESS. 56 9 219-25
PY  - 77
LA  - ENG
CC  - CA051029A

-3-
AN  - CA08726204096S
TI  - THE NATURE OF COAL LIQUEFACTION PRODUCTS
AU  - STERNBERG, HEINZ W.; RAYMOND, RAPHAEL; SCHWEIGHARDT, FRANK K.
OS  - LAWRENCE LIVERMORE LAB., UNIV. CALIFORNIA, LIVERMORE
SO  - PREPR., DIV. PET. CHEM., AM. CHEM. SOC. 21 1 198-209
PY  - 76
LA  - ENG
CC  - CA051029A

-4-
AN  - CA08726204095R
TI  - FRIEDEL-CRAFTS ALKYLATION OF BITUMINOUS COALS
AU  - HODEK, W.; MEYER, F.; KOLLING, G.
OS  - BERGBAU-FORSCH. G.M.B.H., ESSEN
SO  - ACS SYMP. SER. 55 IND. LAB. ALKYLATIONS 408-16
PY  - 77
LA  - ENG
CC  - CA051029A

-5-
AN  - CA08726204093P
TI  - CATALYTIC HYDROGENATION OF COAL (THE RAG/STEAG LARGE-SCALE PILOT
      PLANT)
AU  - WOLOWSKI, ECKARD; LANGHOFF, JOSEF; SCHLUPP, KARL F.; WIEN, HELMUT
OS  - GES. VERGASUNG UND VERFLUESSIGUNG VON STEINKOHLE M.B.H., ESSEN
SO  - COMPEND. - DTSCH. GES. MINERALOELWISS. KOHLECHEM. 76-77 1 373-83
PY  - 76
LA  - GER
CC  - CA051028A

-6-
AN  - CA08726204091M
TI  - COMPARATIVE EVALUATION OF HIGH- AND LOW-TEMPERATURE GAS CLEANING
      FOR COAL GASIFICATION-COMBINED CYCLE POWER SYSTEMS
AU  - JONES, C.H.; DONOHUE, J.M.
OS  - STONE AND WEBSTER ENG. CORP., NEW YORK
SO  - REPORT EPRI-AF-416, 151 PP.
PY  - 77
AV  - NTIS
LA  - ENG
CC  - CA051026A

-7-
AN  - CA08726204090K
TI  - PYROLYSIS OF LARGE COAL BLOCKS:  IMPLICATIONS OF HEAT AND MASS
      TRANSPORT EFFECTS FOR IN SITU GASIFICATION
AU  - WESTMORELAND, P.R.; FORRESTER, R.C., III
OS  - OAK RIDGE NATL. LAB., OAK RIDGE
SO  - REPORT CONF-770301-13, 11 PP.
PY  - 76
AV  - NTIS
LA  - ENG
CC  - CA051026A
```

(Figure 14—continued)

```
-8-
AN   - CA08726204088R
TI   - DESIGN OF AUTOTHERMIC COAL GASIFICATION PLANTS FOR DIFFERENT
       PRODUCT GAS COMPOSITIONS
AU   - NEUMANN, KLAUS KURT; NITSCHKE, EBERHARD
OS   - FRIEDRICH UHDE G.M.B.H., DORTMUND
SO   - COMPEND. - DTSCH. GES. MINERALOELWISS. KOHLECHEM. 76-77 1 328-47
PY   - 76
LA   - GER
CC   - CA051026A

-9-
AN   - CA08726204087Q
TI   - SLAGGING GASIFIER AIMS FOR SNG MARKET
AU   - SAVAGE, PETER R.
SO   - CHEM. ENG. (N. Y.) 84 19 108-9
PY   - 77
LA   - ENG
CC   - CA051026A

-10-
AN   - CA08726204086P
TI   - ALTERNATE FUELS FROM COAL
AU   - MILLS, G.ALEX
OS   - MATER. EXPLORATORY RES., ERDA, WASHINGTON
SO   - CHEMTECH 7 7 418-23
PY   - 77
LA   - ENG
CC   - CA051026A

-11-
AN   - CA08726204085N
TI   - COMBUSTION PROCESSES IN IN SITU COAL GASIFICATION:  PHENOMENA,
       CONCEPTUAL MODELS AND RESEARCH STATUS.  PART I.  OVERVIEW AND
       CONTINUUM WAVE DESCRIPTIONS
AU   - CORLETT, R.C.; BRANDENBURG, C.F.
OS   - DEP. MECHAN. ENG., UNIV. WASHINGTON, SEATTLE
SO   - WEST. STATES SECT., COMBUST. INST., (PAP.) WSS/CI 77-3, 23 PP.
PY   - 77
LA   - ENG
CC   - CA051026A

-12-
AN   - CA08726204063D
TI   - ELECTRICAL CONDUCTIVITY OF COAL AND COAL CHAR
AU   - DUBA, ALFRED G.
OS   - LAWRENCE LIVERMORE LAB., UNIV. CALIFORNIA, LIVERMORE
SO   - FUEL 56 4 441-3
PY   - 77
LA   - ENG
CC   - CA051020A

-13-
AN   - CA08726203972N
TI   - VHTR ENGINEERING DESIGN STUDY:  INTERMEDIATE HEAT EXCHANGER
       PROGRAM
AOS  - GENERAL ELECTRIC CO.
OS   - ENERGY SYST. TECHNOL. DIV., FAIRFIELD
SO   - REPORT COO-2841-1, 436 PP.
PY   - 76
AV   - NTIS
LA   - ENG
CC   - CA051005A
```

(Figure 14—continued)

```
-14-
AN  - CA08726203891K
TI  - INDUSTRIAL GAS PRODUCTION BASED ON KOPPERS-TOTZEK COAL
      GASIFICATION PROCESS
AU  - FRANZEN, JOHANNES E.; GOEKE, EBERHARD K.
OS  - DIV. RES. DEV., KRUPP-KOPPERS G.M.B.H., ESSEN
SO  - USOS CARBON SIDER.:  ABASTECIMIENTO TECNOL., TRAB. CONGR.
      ILAFA-CARBON 333-41
PY  - 76
PUB - INST. LATINOAM. FIERRO ACERO SANTIAGO, CHILE
LA  - SPAN
CC  - CA051000A

-15-
AN  - CA08726203890J
TI  - USE OF COAL GASIFICATION IN THE DIRECT REDUCTION MIDREX PROCESS
AU  - CARINCI, GABRIELE G.; MEISSNER, DAVID C.
OS  - MIDREX CORP., PITTSBURGH
SO  - USOS CARBON SIDER.:  ABASTECIMIENTO TECNOL., TRAB. CONGR.
      ILAFA-CARBON 305-12
PY  - 76
PUB - INST. LATINOAM. FIERRO ACERO SANTIAGO, CHILE
LA  - SPAN
CC  - CA051000A

-16-
AN  - CA08726203865E
TI  - THE MELCHETT LECTURE:  COAL TO BURN
AU  - CALLCOTT, T.G.
OS  - CENT. RES. LAB., BROKEN HILL PROPR. CO. LTD., NEWCASTLE
SO  - J. INST. FUEL 50 403 87-106
PY  - 77
LA  - ENG
CC  - CA051000A

-17-
AN  - CA08726203863C
TI  - COAL
AU  - RINCON, JOSE M.
OS  - FAC. CIENC., UNIV. NAC. COLOMBIA, BOGOTA
SO  - QUIM. IND. (BOGOTA) 8 4 20-2
PY  - 76
LA  - SPAN
CC  - CA051000A

-18-
AN  - CA08726203857D
TI  - SIMILARITIES AND DISSIMILARITIES IN PETROLEUM HEAVY OIL AND COAL
      - COAL LIQUIFACTION
AU  - SANADA, YUZO
OS  - NATL. RES. INST. POLLUT. RESOUR., JAITAMA
SO  - SEKIYU GAKKAI SHI 17 10 835-41
PY  - 74
LA  - JAPAN
CC  - CA051000A
```

Figure 14 Example of on-line search of* Chemical Abstracts. *The user has requested all titles (excluding patents) dealing with coal liquefication and gasification that have appeared since the twenty-sixth file update of 1977. This is based on the System Development Corporation file.

needs to conduct a conventional manual search. Thus on-line services are complementary to conventional holdings and search procedures.

A thorough knowledge of the data base being searched is as essential for on-line searching as it is for manual searching. The user needs to know as much as possible about the data base, including scope, kind of indexing used, and any special features or limitations.

Additionally, the user needs a thorough knowledge of the on-line system being used. This does not mean that the user needs to become a computer programmer or electronics specialist, but it does mean that a basic requirement is study and knowledge of the manuals that suppliers of on-line systems make available. Further competence can be gained by attending training sessions held by suppliers at central locations, or as an option, at the user's own facility. Expertise in use of systems can be further honed by on-line experience in "real-life" situations. A number of months may elapse before a user can be considered "expert." Further, on-line suppliers frequently make improvements and other changes in modes of operation. Secondary sources that constitute input to the system also change. The user needs to keep up with these changes by perusal of the newsletters that suppliers issue periodically.

Expertise with any one system does not assure competence in use of other systems, although all have some fundamental similarities. For this and other reasons, user organizations should train more than one person, each of whom specializes in one or two systems.

Basic typing skill is useful but not essential. This helps because communication with the computer is by typewriterlike keyboard. Ability to operate this equipment quickly and accurately saves time and money.

Most laboratory chemists have their on-line searches conducted for them by chemical information specialists or chemistry librarians. The role of the laboratory chemist is to provide the person who does the searching with complete understanding of what is wanted and to help evaluate search strategy and output. Best results come from close partnership.

COSTS

There are two basic components of on-line system use costs. The first is based on the time the user is connected to the central computer. This unit, often referred to as "connect time," can vary from about $20 per hour to over $100 per hour, depending on the data base and system being used. As a rough rule of thumb, about $1 per minute, or $60 per hour can be considered typical, based on 1977 costs.

An additional related cost, which applies only to some data bases, is a basic subscription fee that the user organization must pay to have access to the data base.

The second basic component of cost is purchase, rental, or lease of the terminal required to interface with the central computers. These costs vary widely, depending on manufacturer and special features selected. They can be as low as about $2000 (1977 cost for purchase of a quality unit).

Examples of other (relatively minor) factors which the user will want to take into account include:

a Additional equipment that may be needed to "couple" the terminal to the computer;

b Any use of special telephone network lines (average cost $8 per hour of "connect time");

c Any savings from requesting an "off-line" print of abstracts or citations at the computer centers and subsequent mailing of this output to the user—this approach can be advantageous when the user wishes to scan a larger number of references than usual.

SEARCH STRATEGY ON-LINE

For most effective search results, the plan of attack must be developed on paper or blackboard before going on-line. Search objectives, basic logic, and search terms should be specified clearly. Certain of the search aids, such as word-frequency-count lists, are available on microform specifically to help on-line searchers in their planning. Once on-line, the user has at his disposal additional searching aids, some of which are not available manually.

The user will usually scan a small sample of initial results in an on-line search. If necessary, search strategy can then be quickly and easily changed while the user is still on-line to achieve the most acceptable results. The capability of revising search strategy easily is one of the most attractive features of on-line searching. It is a major reason why this relatively new tool is now used so widely and effectively.

In the use of chemical indexes, especially those searched on-line, it is important to look under as many variations as possible. In some cases, thesauri or related types of guides are available. These suggest or specify some variants that should be examined. For *CA*, a basic tool for this purpose is the *Index Guide* (see p. 52), and another aid is the list of abbreviations found in the *Directions for Abstractors* (8-1).

One example of use of as many variations as possible in on-line searching occurs when the chemist is looking for references on ways of making a chemical. Some variations which need to be considered include:

1 a. Manuf. d. Manufacture
b. Manufd. e. Manufactured
c. Manufg. f. Manufacturing

These usually denote references to methods of making chemicals on a commercial or plant scale.

Probably the most certain way that the chemist or engineer has of determining that a reference containing one of the above variations *may* pertain to commercial manufacture of a chemical is identification of that reference as a patent (see Chapter 11 on patents). Much patent literature has at least the potential goal of commercial manufacture. In *CA*, association of an index entry with a patent is indicated by the symbol "P."

2 a. Prep.　　e. Prepared
b. Prepd.　　f. Preparing
c. Prepg.　　g. Preparation
d. Prepare　　h. Synthesis

These almost invariably refer to methods of making chemicals on a laboratory or bench scale, but, again, when the reference is identified as a patent, commercial scale manufacture is the likely object.

3 Formation

This usually signifies work in which the chemical occurs as a minor byproduct, or only incidentally to preparation of another compound which is the prime object of the investigation.

Other information pertinent to the making of a chemical could be found under such keywords or subject entries as:

4 Fermentation (and related terms)
5 Isolation (and related terms)
6 Process (and related terms)
7 Production (and related terms)
8 Purification (and related terms)
9 Recovery (and related terms)

But even the inclusion of all terms 1 through 9, as noted above may not be enough. An article may relate to preparation of a compound without this information appearing in either the title or words used to index the article. For example, when a physical property of a substance is the principal object of the investigation, the article may include a report on preparation of a pure sample of the substance.

Also, the compound may be prepared or formed as a minor intermediate, and reported on in an article dealing primarily with preparation of another compound, but not indexed. This depends on the policy of the information tool.

The key used by indexers for some of the most significant information tools is emphasis placed on the subject by the author of the original document.

As another example, consider the case of a chemical engineer concerned with selection of materials of construction to be used with a specific product. In addition to "materials of construction" as an entry in an on-line search, other entries to be searched could include:

1 Anticorrosive (and variations)
2 Compatibility (and variations)
3 Corrosion (and variations)
4 Inhibition (and variations)
5 Noncorrosive (and variations)
6 Prevention (and variations)
7 Protection (and variations)
8 Resistance (and variations)
9 Specific materials, such as a type of steel
10 Stabilization (and variations)

Or consider the case of an analytical chemist searching for references on gases evolved on decomposition of a polymer. If this search were conducted on-line, an array such as listed in Figure 15 could be constructed. Note that this array takes into account variant terms as well as abbreviations, all of which

decompd.	decomposing	dissocd.	dissociating	oxidn.
decompg.	decomposition	dissocg.	dissocn.	pyrolysis
decompn.	degradation	dissociate	heat	temp.
decompose	degrdn.	dissociated	hot	temperature
decomposed	dissoc.	dissociation	oxidation	thermal

Figure 15 Examples of term variations used in on-line searching.

would need to be searched for to ensure that the search is as complete as possible.

The chemist must also often include in his search variations in names of chemicals. A simple case is polyvinyl chloride which might be found entered as any one or all of the following examples:

Ethene, chloro-, polymers, homopolymer
PVC
Polyvinychloride (one word)
Poly(vinyl chloride)
Polyvinyl chloride (two words)
Vinyl
Vinyl compounds, polymers
Vinyls

Note the inclusion of both singular and plural variations of the same word. Also note that the above list is not necessarily all inclusive; other variants may need to be used.

Chapter 3 on Search Strategy contains some suggestions that could be of specific interest to planning of on-line investigations.

INFORMATION CENTERS

There are a number of large "information centers" in the United States and other major industrial countries (for a partial list, see Figure 16). These centers provide a wide variety of current awareness and retrospective searching services.

The services are usually computer based. A few provide primarily on-line service. Most are, however, "batch" or "off-line." Input (questions) is transmitted by mail or ordinary voice telephone line, and output (answers) or search results are usually transmitted to users by mail. Response time is usually overnight at best, but more typically several days, depending on the mails.

Centers that provide off-line or batch service continue to be an important factor, despite the increasing popularity of on-line services. Information centers offer such benefits as one or more of the following:

1 Providing expert assistance in correct structuring of inquiries, that is, concise and comprehensive statement of questions.

2 Assisting in developing search strategy that will be most likely to yield desired results.

3 Selecting those files or sources that are most likely to yield fruitful output.

4 Organizing, screening, and evaluating for pertinent output or results.

5 Supplementing results with critique and data, when the centers have on their staffs persons who possess the required expertise.

6 Using computer programs in some cases more advanced and "powerful" than on-line services.

7 Output in a variety of forms, such 8½ × 11″ paper or index card form.

8 Eliminating of need for user to rent or buy special equipment (such as interactive terminals) to use this kind of service.

9 Eliminating of need for user to learn special search languages to communicate with the computer.

10 Assisting in obtaining copies of pertinent documents identified.

11 Referrals to, and assisting with, making contact with experts in outside organizations who may have knowledge on the question at hand.

Some persons question whether such information centers will survive in the wake of the tidal wave of on-line services. But this author believes that there will be a long-term need for off-line information centers, such as those just described, especially for smaller organizations and for independent chemists such as consultants.

Such factors as economics, availability of user and computer time, and computer program capability will determine the relative popularity of off- and on-line services and information centers in the future.

INFORMATION "BROKERS"

There are a number of independent information "brokers" who can conduct, or arrange to have conducted, both on- and off-line computer-based searches of the literature.

These brokers often offer a wide variety of information services (including manual searches) and can be of particular value to smaller organizations or to "overloaded" larger organizations.

The best source for obtaining the names and locations of many of the most capable of the brokers is the Information Industry Association, 316 Pennsylvania Ave., S.E., Suite 502, Washington, D.C. 20003.

Chemical Abstracts Service (CAS) has licensed a number of organizations to use one or more CAS files from which these organizations derive marketable services for their respective customers. These information centers provide a variety of services, among which are current awareness services and retrospective searches. Centers often provide an economical means for individuals and organizations to obtain needed information immediately while minimizing cost since charges are generally based on the specific service rendered.

Individual centers operate differently, providing different services and using different CAS files. Many offer services based on files from other suppliers. Inquiries regarding operation of a specific center should be directed to the center.

This list is believed to be up-to-date as of February, 1977. It was compiled by CAS and is reprinted here through the courtesy of CAS. Those U.S. centers which are primarily vendors of on-line services are listed on p. 80 and are not repeated here.

INFORMATION CENTERS WITHIN THE UNITED STATES

CENTRAL ABSTRACTING AND INDEXING SERVICE
American Petroleum Institute
Manager
275 Madison Avenue
New York, New York 10016
Phone: (212) 685-6253

CHEMICAL INFORMATION CENTER
Director
Department of Chemistry
Indiana University
Bloomington, Indiana 47401
Phone: (812) 337-9452/3

GEORGIA INFORMATION DISSEMINATION CENTER
Manager
Office for Computing Activities
Boyd Graduate Studies Building
University of Georgia
Athens, Georgia 30602
Phone: (404) 542-3106

GTE LABORATORIES, INCORPORATED
Technical Information Center
40 Sylvan Road
Waltham, Massachusetts 02154
Phone: (617) 890-8460

IIT RESEARCH INSTITUTE
Manager, Computer Search Center
10 West 35th Street
Chicago, Illinois 60616
Phone: (312) 567-4000

THE INSTITUTE OF PAPER CHEMISTRY
Director, Division of Information Services
1043 East South River Street
Appleton, Wisconsin 54911
Phone: (414) 734-9251

KNOWLEDGE AVAILABILITY SYSTEMS CENTER
Director
300 LIS Building
University of Pittsburgh
Pittsburgh, Pennsylvania 15260

***Figure 16** Information centers that provide services based on files licensed from Chemical Abstracts Service and (in many cases) from other sources.*

Phone: (412) 624-5212
TWX: (710) 664-3060

MECHANIZED INFORMATION CENTER
The Ohio State University
10 Lazenby Hall
1827 Neil Avenue
Columbus, Ohio 43210
Phone: (614) 422-3480

MERIT INFORMATION CENTER, CHEMISTRY DIVISION
Director
Department of Chemistry
The University of Michigan
Ann Arbor, Michigan 48104
Phone: (313) 764-7362

NEW ENGLAND RESEARCH APPLICATION CENTER
University of Connecticut
New England Research Center
Mansfield Professional Park
Storrs, Connecticut 06268
Phone: (203) 486-4533
TWX: (710) 420-0571

UNIVERSITY OF NORTH CAROLINA
University Librarian
Wilson Library
Chapel Hill, North Carolina 27514
Phone: (919) 933-1301

UNIVERSITY OF CALIFORNIA
Computerized Information Services
32 Powell Library Building
Los Angeles, California 90024
Phone: (213) 825-1573

UNIVERSITY OF IOWA
Head Information Retrieval Services
University Computer Center
Iowa City, Iowa 52242
Phone: (319) 353-3170

*INSTITUTE FOR SCIENTIFIC INFORMATION
325 Chestnut Street
Philadelphia, Pennsylvania 19106
Phone: (215) 923-3300

*(This Center does not appear in the CAS list; it was added by the author. It provides services based on its own data bases only.)

INFORMATION CENTERS OUTSIDE CONTINENTAL UNITED STATES

ARGENTINA

CENTRO DE DOCUMENTACION CIENTIFICA DEL CONICET
Jefe de Dpto.
Moreno 431/433
Buenos Aires, Argentina
Telex: 122414AR CEDOC

AUSTRALIA

CSIRO
Central Information Services
314 Albert Street
East Melbourne, Victoria, Australia 3002
Phone: 419-1333

(Figure 16—continued)

AUSTRIA

Rechenzentrum Graz
Chemie Informationsdienst Graz
Steyrergasse 17
A-8010 Graz, Austria
Phone: 73 5 21, 87 3 63, Direktion 83 4 04
Telex: 03/1265RZ GRAZ

BELGIUM

ROYAL LIBRARY OF BELGIUM
Director, CNDST
Bd. de l'Empereur 4
B-1000 Brussels, Belgium
Phone: (02) 513.61.80 — Telex: 21157

CANADA

CANADA INSTITUTE FOR SCIENTIFIC AND TECHNICAL INFORMATION (CISTI)
National Research Council of Canada
Head, Tape Services
Ottawa KIA OS2, Ontario, Canada
Phone: (613) 993-1210 Telex: 053-3115
(provides on-line service)

CZECHOSLOVAKIA

Central Office for Scientific, Technical and Economical Information (UVTEI)
Division Central Technical Basis (UTZ)
Konviktska 5
113 57 Prague 1, Czechoslovakia
Telex: Praha 012114

DENMARK

DANMARKS TEKNISKE BIBLIOTEK
Anker Engelunds vej 1
DK 2800 Lyngby, Denmark
Phone: Copenhagen 88 30 88 Telex: 15048

ENGLAND

UNITED KINGDOM CHEMICAL INFORMATION SERVICE
Manager, Service Department
The University
Nottingham NG7 2RD, United Kingdom
Phone: 0602-57411/5 Telex: 37488

FRANCE

CENTRE NATIONAL de L'INFORMATION CHIMIQUE
88, Avenue Kleber
75116 Paris, France
Phone: 553-65-19

EUROPEAN SPACE AGENCY
8-10, Rue Mario Nikis
75738 Paris-Cedex 15, France
Tel: 5675578 Telex: 202746

GERMANY

CHEMIE-INFORMATION UND -DOKUMENTATION BERLIN
Chefredakteur
Postfach 12 60 50
Steinplatz 2
1000 Berlin 12, Germany
Phone: (030) 31 05 81

HUNGARY

MAGY GYOGYSZERIPARI-EGYESULES
Es-Gyogyszerkutate Intezet
Postafiok 15
1553 Budapest, Hungary

VESZPREMI VEGYIPARI EGYETEM KOZPONTI KONYVTARA
Central Library of the Veszprem University of Chemical Engineering

(Figure 16—continued)

Director of the Library
Schonherz Z.u.10.
8201 Veszprem, Hungary
Phone: 12-550 Telex: 32-397/Hungary

ISRAEL

NATIONAL CENTER OF SCIENTIFIC AND TECHNOLOGICAL INFORMATION
Head, Department of Users Liason
Hachasmonaim Street 84
P. O. Box 20125
Tel Aviv 61 200, Israel
Phone: 03-288266 — Telex 03-2332

ITALY

EUROPEAN SPACE AGENCY
Space Documentation Service
Via Galileo Galilei
00044 Frascati, Italy
Phone: 9422401 Telex: 61637 ESRINROM
(Provides on-line service)

JAPAN

THE JAPAN INFORMATION CENTER OF SCIENCE AND TECHNOLOGY
Head, Literature Search Section
5-2, 2-Tyome, Nagatatyo, Tiyoda-ku
Tokyo 100, Japan
Phone: Tokyo 581-6411
(Provides on-line and off-line services)

KINOKUNIYA BOOK-STORE COMPANY, LIMITED
Manager, ASK System Department
4-10-13 Kitashinjuku
Shinjuku-ku
Tokyo 160, Japan
Phone: 354-0131 Area Code -03

KOREA

KOREA SCIENTIFIC AND TECHNOLOGICAL INFORMATION CENTER (KORSTIC)
Director General
206-9, Cheongryangri-dong, Dongdaemun -ku
C.P.O. Box 1229
Seoul, Korea
Tel: 96-6501-6

NETHERLANDS

CENTRE FOR TECHNICAL AND SCIENTIFIC INFORMATION AND DOCUMENTATION TNO (CID-TNO)
P.O. Box 36
2600 AA, Delft, The Netherlands
Tel: (15) 569330 Telex: 31453 zptno nl.

EUROPEAN SPACE AGENCY
European Space Technology Centre
Domeinweg
Noordwijk, The Netherlands
Tel: (1719) 86555 Telex: 31698

POLAND

WARSAW TECHNICAL UNIVERSITY
Director
Central Library
Politechnilsi Warszawskiej
Plac Jednosci Robotniczej 1
Warsaw, Poland

TECHNICAL UNIVERSITY OF WROCLAW
Main Library and Scientific Information Centre
ul. Wybrzeze Wyspianskiego 27
50-370 Wroclaw, Poland
Phone: 20 23 05 Telex: 03 42 54 Polit PL

(Figure 16—continued)

SOUTH AFRICA

SOUTH AFRICAN COUNCIL FOR SCIENTIFIC AND INDUSTRIAL RESEARCH
Centre for Scientific & Technical Information
Head, SASDI
P. O. Box 395
Pretoria
0001 Republic of South Africa
Phone: (012) 74-9111 Telex: 3-630

SPAIN

CENTRO DE INFORMACION Y DOCUMENTACION
Director
Joaquin Costa, 22
Madrid 6, Spain
Telex: 831-22628 CIDMD E

SWEDEN

SWEDISH COUNCIL FOR SCIENTIFIC INFORMATION AND DOCUMENTATION
Medicinska Informationscentralen, MIC
Head, Section of Chemical Documentation
Karolinska Institutet
Fack S-104 01 Stockholm 60, Sweden
Phone: 08-23 2270 Telex: 17178 Karolin S

SWEDISH COUNCIL FOR SCIENTIFIC INFORMATION AND DOCUMENTATION
The Royal Institute of Technology Library
Head, Information and Documentation Center, IDC
S-100 44 Stockholm 70, Sweden
Phone: 08-23 65 20 Telex: 103 89 KTHB Stockholm

(Figure 16—continued)

9 REVIEWS

Comprehensive and evaluative reviews can minimize or obviate the need for the chemist to do any further literature searching on the subject in question. Reviews save time and help provide perspective.

Some questions which help the user evaluate the quality of a review include:

1 Is the author of the review a known expert in the field?

2 Is there a clear statement of scope and purpose?

3 What is the time frame covered? How up-to-date are the most recent sources included, and how far back in time does the author go? The more comprehensive the better.

4 Is the author critical or evaluative? Does he identify what he believes is the most important and valid data?

5 Is the material well organized and presented?

6 Is there a complete bibliography with all sources clearly identified?

The best known of the review journals is *Chemical Reviews (CR)*. Initiated in 1925, and now published monthly by the American Chemical Society, *CR* contains critical, comprehensive reviews by experts in the field, with emphasis on the more theoretical aspects of chemistry.

A cumulative index to the volumes for 1925–1960 appears in Volume 60 (1960), and Volume 70 (1970) contains an index to the volumes for 1961–1970. These collective indexes are in addition to conventional annual indexes.

Other significant review tools include these examples:

- *Annual Reports on the Progress of Chemistry* (published annually since 1904 by The Chemical Society). Section A deals with physical and inorganic chemistry and Section B with organic chemistry. Emphasis is on "pure chemistry."

- *Chemical Society Reviews* (published quarterly since 1972 by The Chemical Society.)

- *Reports on the Progress of Applied Chemistry* (published annually for more than 60 years by the Society of Chemical Industry.) The title is self-explanatory. (The author understands that this will soon be replaced by a new series of critical reviews.)

- *Russian Chemical Reviews* (English translation of *Uspekhi Khimii*)

- *Specialist Periodical Reports* (This series was launched by The Chemical Society in 1967 to provide "systematic and comprehensive review coverage of the progress in major areas of chemical research." The series has now reached about 35 titles, some of which are published annually and others as biennial volumes.) Examples of some titles include:

Aliphatic, alicyclic and saturated heterocyclic chemistry

Aliphatic chemistry

Alicyclic chemistry

The alkaloids

Amino-acids, peptides and proteins

Aromatic and heteroaromatic chemistry

Biosynthesis

Carbohydrate chemistry

Chemical thermodynamics

Colloid science

Dielectric and related molecular processes

Electrochemistry

Electron spin resonance

Electronic structure and magnetism of inorganic compounds

Environmental chemistry

Fluorocarbon and related chemistry

Foreign compound metabolism in animals

Inorganic chemistry of the main group elements

Inorganic chemistry of the transition elements

Inorganic reaction mechanisms

- Mass spectrometry
- Molecular spectroscopy
- Molecular structure by diffraction methods
- Nuclear magnetic resonance
- Organic compounds of sulphur, selenium and tellurium
- Organometallic chemistry
- Organophosphorous chemistry
- Photochemistry
- Radiochemistry
- Reaction kinetics
- Saturated heterocyclic chemistry
- Spectroscopic properties of inorganic and organometallic compounds
- Statistical mechanics
- Surface and defect properties of solids
- Terpenoids and steroids
- Theoretical chemistry

Additionally, the CRC Press, 18901 Cranwood Parkway, Cleveland, OH, 44128, has initiated a series of review journals which include the following examples:

CRC Critical Reviews in Analytical Chemistry
CRC Critical Reviews in Biochemistry
CRC Critical Reviews in Toxicology

Chemical Abstracts (CA) editors have recognized the importance of reviews by designating these with the symbol "R" in the indexes to *CA*. Many chemists examine abstracts of reviews before they look at abstracts for other index entries, because a good review can provide all that is immediately needed on a topic.

The United Kingdom Chemical Information Service at the University of Nottingham publishes a comprehensive index to review articles appearing in *CA*. This index, known as *CA Reviews Index (CARI)*, began in 1975 and is produced twice a year. Each issue corresponds to a volume of *CA*.

An *Index to Scientific Reviews (ISR)* published by the

Institute for Scientific Information (see p. 73), helps locate more than 23,000 review articles published each year in chemistry and other scientific disciplines. *ISR* is issued semiannually and cumulated annually.

Not all reviews appear in journals and books specifically devoted to reviews. A good review can appear in almost any chemical journal or other information source.

10 ENCYCLOPEDIAS AND OTHER MAJOR REFERENCE BOOKS

Journals and patents almost always contain the latest information, and that is frequently what chemists need.

The chemist, however, cannot afford to ignore books. Books are usually the best starting point for searching the chemical literature. They can often provide quick, straightforward answers that could otherwise take many hours to locate in widely scattered journal articles and patents on the subject. Books provide the foundation from which to launch any further search of the literature required.

When the literature on a chemical or reaction is voluminous, use of an appropriate book is highly recommended to help sort out what is important and vital from what is not. Some "advanced" books are evaluative or critical and can help minimize further searching. Review journals and other review sources (see Chapter 9) perform a similar function.

The chemist will find some older books, seemingly obsolete, of value. As noted later in this chapter, older editions may contain some information not included in more recent editions, and this may be precisely what the chemist needs. If a search is being made to see if an idea being considered for a patent is novel, consultation of such older books is mandatory.

Various kinds of books are referred to throughout this volume. This chapter focuses on some of the major reference works found in most good chemistry libraries.

First, there is a discussion of significant encyclopedias in chemistry and chemical engineering. This is followed by brief remarks on the German "Handbuch" concept and a review of some important reference works in organic and inorganic chemistry.

ENCYCLOPEDIAS—INTRODUCTORY REMARKS

The expert or specialist in a field will usually find treatment of this field by encyclopedias too general and elementary. Additionally, encyclopedias are quickly dated. By the time the last volume is completed, initial volumes may be obsolete.

But most chemists and engineers beginning work on a new project about which they know little or nothing will find that the major chemical encyclopedias are good starting points. For further details the user can then consult the monographs, patents and articles cited in the encyclopedias, and more recent sources as required. The relative conciseness which is a feature of encyclopedias is usually a plus because it gives the chemist a quick overview of a field.

Many current encyclopedia articles, however, probably as a matter of editorial policy, do not cover the recent process technology and economics for important commercial products with the depth of analysis and detail some users require. The special tools which provide this kind of information are described in Chapter 15, p. 230.

The principal encyclopedias of interest to chemical engineers and chemists include:

1 *Encyclopedia of Chemical Technology* (commonly referred to as *Kirk-Othmer* after the names of the original editors).

2 *Ullmanns Encyclopädie der Technischen Chemie.*

3 *Encyclopedia of Chemical Processing and Design* (edited by J. J. McKetta).

4 *Encyclopedia of Polymer Science and Technology—Plastics, Resins, Rubbers, Fibers* (frequently referred to by the name of its Editor, Dr. Herman F. Mark).

Each is described in the paragraphs which follow.

Specific Encyclopedias

1. *Kirk–Othmer* (10–1). Most chemists and chemical engineers agree that, with the exception of *Chemical Abstracts*, *Kirk–Othmer (KO)* is the most indispensable tool in any chemistry or chemical engineering library, at least in the United States. There are already two complete editions of *KO*, and a third edition is in preparation, as is noted later.

The initial volume of the first edition of *Kirk–Othmer* was published in 1947, and the final volumes (in this case *Supplements*) were published in 1960.

The first volume of the second edition of *Kirk—Othmer* appeared in 1962, and the *Supplement* and *Index* were published in 1971 and 1972. Clearly, then, some material, especially in the older volumes, is obsolete.

But the first edition of *KO* remains useful and is still consulted. It contains some information of value that does not appear in the second edition.

Processes, products, and concepts once believed obsolete may again become interesting because of changing economics, raw materials shortages, or environmental reasons. This is a principal reason why older editions of encyclopedias such as *KO* can maintain a surprisingly long useful life span.

When the third edition is completed, the first and second editions will probably continue to be useful for the same reason.

In both the first and second editions of *KO*, principal, but by no means total, emphasis is placed on the more applied aspects—the application of chemistry and chemical engineering to industrially important concepts, products, processes, and uses.

Most articles in *KO* provide good coverage of significant aspects of the topics covered. Numerous references to patent, journal, and other literature permit chemists and engineers using *KO* to pursue fields of interest in more depth.

KO is not all inclusive and is characterized by the uneven treatment that is inevitable in an encyclopedia in any field. Also, if an article on a chemical is written by an employee of a manufacturer of that product, the article may provide strong coverage for properties and uses but weak coverage of manufacturing details. This approach helps users of the product but is otherwise incomplete.

KO is an American encyclopedia, written mostly by American authors for American chemists and engineers. Although there are discussions of, and references to, European and Japanese technology, primary emphasis is on United States practice. Omission of foreign practice is a shortcoming in some fields.

As mentioned, a new edition of *KO* (the third) under the editorship of Dr. Martin Grayson, is under way as this book goes to press. The initial volume is scheduled to appear in 1978, and the target date for completion is 1983. According to present plans, there will be 25 volumes, published at a rate of approximately four volumes per year.

This author hopes that the third edition will provide better coverage of European, Japanese, and other foreign developments, especially in high-technology fields of significance.

2. *Ullmanns* (10-2). The German language counterpart of *KO* is *Ullmann's Encyklopaedie der Technischen Chemie*. Publication of the fourth edition of this important work began in 1972. Plans call for the issuance of two volumes per year.

According to the publisher, "Volumes 1–6 (*Thematic Section—General Principles & Methodology*) will constitute a complete, up-to-date account of general principles and methodology (e.g., thermodynamics, process engineering and development); physical-chemical analysis methods; and measurement and control, pollution abatement, and works safety." The following 18 volumes will consist of an alphabetical compilation of articles on key subjects. Every third volume of the 18 alphabetical volumes will contain a cumulative subject index. Volume 25 will be a cumulative subject index to the complete work.

As compared to *KO*, *Ullmanns* has both advantages and disadvantages. Although coverage of chemistry is about the same, *Ullmanns* seems to cover process technology in more detail, at least compared to the second edition of *KO*.

If previous editions of *Ullmanns* can be taken as an indication, the industrial chemist and engineer can be assured of full and accurate treatment of important topics of interest. The literature citations accompanying each article can be expected to be of special value, with appropriate emphasis given to pertinent sources, especially patents. Coverage of developments in Europe, and especially West Germany, surpasses that of the second edition of *KO*, as might be anticipated.

The principal disadvantage of *Ullmanns*, as far as the typical English-speaking and reading chemist or engineer is concerned, is that the text of this work is in German. An attempt is made to bridge the language barrier by providing both German- and English-language tables of contents and indexes.

Nevertheless, some lessening of speed and comprehension for chemists who are not thoroughly familiar with German chemical terminology and technology is inevitable. Additionally, and equally important, practices and conventions described may not correspond with those of the United States and other countries in some cases. Yet *Ullmanns* is clearly a key source well worth consulting.

3. *McKetta* (10–3). The initial volumes of a totally new encyclopedia, targeted primarily at chemical engineers, are now (1977) being published. This is the *Encyclopedia of*

Chemical Processing and Design, edited by Dr. John J. McKetta, Professor, Department of Chemical Engineering, University of Texas, Austin. The stated purpose of this publication is to provide a modern, up-to-date encyclopedia detailing current developments in the field of chemical technology.

According to the publisher, "the encyclopedia rigorously covers chemical processes, methods, practices and standards used in the chemical industry. Although most articles are written by American chemical engineers, scientists and technologists, no important foreign source has been neglected in making the encyclopedia truly international in scope."

As the title of this work indicates, emphasis is placed on chemical processing and, particularly, design aspects. This encyclopedia is scheduled to contain at least 20 volumes.

A full comparison of *McKetta* with the new editions of *KO* and *Ullmanns* will need to await completion of these three important compilations.

4.*Mark* (10-4). For the chemist or engineer concerned primarily with polymer chemistry, the major encyclopedic source, in lieu of or in addition to those previously mentioned, is the *Encyclopedia of Polymer Science and Technology*.

Which Encyclopedia To Use

In their use of major chemical encyclopedias, chemists and engineers should consider *all* of those mentioned as appropriate, because each can provide a different perspective, may have different or additional information, and because topics not adequately covered in one encyclopedia may be more fully and adequately covered in another. The staggered and overlapping time frames within which the encyclopedias are being issued is another reason for looking at all of those listed.

THE HANDBUCH CONCEPT

The *Handbuch* is a valuable type of publication, significantly different from most handbooks published in the United States. It is a multivolume work, far more extensive in scope, and provides more in-depth coverage, than the typical United States handbook.

Handbuch coverage usually goes back to the "beginnings" of chemistry—far before *Chemical Abstracts* began in 1907—and brings that coverage up to a reasonably current time, as is discussed later in this chapter. Ongoing updating is another feature. For the chemist, this can mean that any needed search of later literature can begin where *Handbuch* coverage ceases.

In many cases, consultation of the *Handbuch* can obviate the need for further literature searching, if the *Handbuch* has the desired information. But it would be a mistake for the chemist to conclude that *Handbuch* coverage is 100% complete and without some errors or omissions.

The two most important examples of the German Handbuch in chemistry are *Beilstein* and *Gmelin*. Both are edited at the Karl Bosch House in Frankfurt, Germany.

The *Handbuch* is an old tradition in German chemistry. In 1817, Leopold Gmelin published his first *Handbuch*, which was a quick success. At the start of the fifth edition (1852), the field of organic chemistry was split off and has been carried on separately as *Beilstein*.

At present, the editors of *Gmelin* are doing a much better job of keeping their volumes more current and more attuned to the needs of the modern industrial chemist and engineer, in addition to those of the academician, that are the editors of *Beilstein*. One reason for this may be because the field of organic chemistry is so large.

Gmelin is now in its eighth edition, and *Beilstein* is in its fourth.

ORGANIC CHEMISTRY—SOME IMPORTANT REFERENCE WORKS

1. Beilsteins Handbuch der Organischen Chemie

Beilstein (10-5) is the basic monumental reference work on organic compounds. It takes its name from Friedrich Karl Beilstein, who published the first edition in 1881–1882.

Many readers of this book will have already heard about, seen, or used this essential tool for organic chemists. It is available in almost every medium-sized or large chemistry library. Further, there is a recent guide (10-6) that describes *Beilstein* and how to use it in some detail. Accordingly, it would be redundant to present an in-depth description of *Beilstein* here and more appropriate to concentrate on some key features.

Beilstein's coverage goes back to the beginnings of organic chemistry. Ongoing supplementing brings it up to about 1960 in many cases as of this writing.

For each chemical compound included, selected available information is presented on, for example:

- Natural occurrence and isolation
- Preparation and manufacture
- Structure of the molecule
- Physical properties alone and in mixture with other compounds
- Chemical properties
- Analytical methods

Input is based on a careful review of the published literature, including journals, patents, reports, and other publications. Each piece of data is accompanied by reference to the publication that is the source of the data.

The editors present the material in highly condensed, telegraphic form. Extensive use is made of abbreviations and symbols. Although this approach may at first seem to make use difficult, most chemists find *Beilstein* easy to read, even with a minimal knowledge of German. Weissbach's *Guide* (10-6) includes a list of English and French equivalents of expressions most frequently used in *Beilstein*.

The table below shows the series and supplements in which *Beilstein* is published.

Series and Supplements	***Approximate Time Period of Literature Coverage***
Main volume (Hauptwerk) up to 1910	up to 1910
Supplement I (Erstes Ergänzungswerk)	1910–1919
Supplement II (Zweites Ergänzungswerk)	1920–1929
Supplement III (Drittes Ergängzungswerk)	1930–1949
Supplement IV (Viertes Ergänzungswerk)	1950–1959

For Volumes 17–27, which cover heterocyclic compounds, Supplements III and IV are combined to cover the years 1930–1959.

Chemists can find compounds in *Beilstein* in three ways: (a) "System" number based on a kind of hierarchy or classification unique to *Beilstein*, (b) subject index, and (c) molecular formula index. Each volume contains its own subject and formula index; collective ("general") indexes are published periodically.

Access by formula is the first choice of this author. Use of the subject index (actually a chemical substance index) is second choice because of the vagaries of chemical nomenclature. The approach by system is a third choice because of the time required to learn the rules involved.

2. Some Other Important Organic Chemistry Reference Works

The principal English-language reference work in organic chemistry is *Rodd's Chemistry of Carbon Compounds* (10-7), the second edition of which began publication in 1964.

This is an outstanding work of broad scope. Although not as comprehensive as *Beilstein*, it is, in many fields, more current and provides an excellent starting point.

Methoden der Organischen Chemie (10-8), also commonly referred to as *Houben-Weyl*, is an extensive multivolume set now in its fourth edition. It gives selected methods for preparation of many organic compounds, in considerable detail.

The *Dictionary of Organic Compounds* (10-9), often referred to as "Heilbron" (Sir Ian Heilbron was chairman of the editorial board until his death in 1959) is a multivolume set which has concise data on many organic compounds. It is not intended as an alternate or replacement for *Beilstein*, which is much more comprehensive. But the chemist will find *Heilbron* useful for quick look-up of such basic data as chemical constitution, a few selected physical and chemical properties, and references to some preparative methods. Supplements bring coverage up to about the mid 1970s.

Organic Syntheses (10-10) is an annual compilation of "satisfactory" and checked laboratory methods for preparing organic compounds. There is a collective volume every 10 years which revises and updates the annual volumes where necessary.

Recent volumes of this series (beginning with Volume 41, 1961) emphasize "widely applicable, model procedures that illustrate important types of reactions; . . . .many of the procedures selected . . . have major significance in the synthetic method, rather than in the product that results. However, preparations of reagents and products of special interest are also included, as in previous volumes." (10-11)

The change in policy is easily explained. When the series was founded by Roger Adams at the end of World War I, there was an urgent need for preparing useful compounds in reasonable quantities, because supplies of organic chemicals from abroad were no longer available. Today, a wide variety of fine organics are commercially available, and reactions can be studied with decreasing starting material quantities because of improved techniques.

Organic Reactions (10-12) are "collections of chapters each devoted to a single reaction, or a definite phase of a reaction, of wide applicability." Emphasis is on the preparative aspects; several detailed procedures are given for each method. Many examples and extensive bibliographies are included. Beginning with Volume 22 (1975), a program of updating earlier reviews was initiated.

Survey of Organic Syntheses (10-13) by Buehler and Pearson is an example of several similar works which "bring together in one volume the principal methods for synthesizing the main types of organic compounds. . . .How the functional group is created from other functional groups is the main concern of [this] work."

Harrison & Harrison's *Compendium of Organic Synthetic Methods* (10-14) is a compilation of published organic functional group transformations. Synthetic methods are presented in the form of reactions with references.

Within *Gmelin* (see below), a supplementary work attempts to build a bridge between classical organic and inorganic chemistry. Thus such volumes on metallo-organic chemistry as the following examples are available:

Chromium—organic bonds
Cobalt—organic bonds
Hafnium—organic bonds
Iron—organic bonds
Nickel—organic bonds
Vanadium—organic bonds
Zirconium—organic bonds

HCl

System-Nr.	Symbol	Element
1		Edelgase
2	H	Wasserstoff
3	O	Sauerstoff
4	N	Stickstoff
5	F	Fluor
6	**Cl**	**Chlor**
7	Br	Brom
8	J	Jod
	At	Astat
9	S	Schwefel
10	Se	Selen
11	Te	Tellur
12	Po	Polonium
13	B	Bor
14	C	Kohlenstoff
15	Si	Silicium
16	P	Phosphor
17	As	Arsen
18	Sb	Antimon
19	Bi	Wismut
20	Li	Lithium
21	Na	Natrium

$CrCl_2$

$ZnCrO_4$

System-Nr.	Symbol	Element
37	In	Indium
38	Tl	Thallium
39	Sc	Scandium
	Y	Yttrium
	La	Lanthan
	Ce–Lu	Lanthanoide
40	Ac	Actinium
41	Ti	Titan
42	Zr	Zirkonium
43	Hf	Hafnium
44	Th	Thorium
45	Ge	Germanium
46	Sn	Zinn
47	Pb	Blei
48	V	Vanadium
49	Nb	Niob
50	Ta	Tantal
51	Pa	Protactinium
52	**Cr**	**Chrom**
53	Mo	Molybdän
54	W	Wolfram
55	U	Uran

$ZnCl_2$

22	K	Kalium
23	NH_4	Ammonium
24	Rb	Rubidium
25	Cs	Caesium
	Fr	Francium
26	Be	Beryllium
27	Mg	Magnesium
28	Ca	Calcium
29	Sr	Strontium
30	Ba	Barium
31	Ra	Radium
32	**Zn**	**Zink**
33	Cd	Cadmium
34	Hg	Quecksilber
35	Al	Aluminium
36	Ga	Gallium

56	Mn	Mangan
57	Ni	Nickel
58	Co	Kobalt
59	Fe	Eisen
60	Cu	Kupfer
61	Ag	Silber
62	Au	Gold
63	Ru	Ruthenium
64	Rh	Rhodium
65	Pd	Palladium
66	Os	Osmium
67	Ir	Iridium
68	Pt	Platin
69	Tc	Technetium
70	Re	Rhenium
71	Np, Pu . . .	Transurane

Figure 17 Gmelin system of elements and compounds. The material under each element number contains all information on the element itself as well as on all compounds with other elements which precede this element in the Gmelin System. For example, zinc (system number 32) as well as all zinc compounds with elements numbered from 1 to 31 are classified under number 32.

INORGANIC CHEMISTRY—SOME IMPORTANT REFERENCE WORKS

1. Gmelins Handbuch der Anorganischen Chemie

The chemist working on elements and inorganic chemistry and compounds, including organometallic compounds, should usually consult *Gmelin* (10-15) as the next step after studying any available encyclopedia article in the field. He may also wish to consult the English language counter-part of *Gmelin* which is *Mellor*, discussed in the next section.

Of these two sources of information on inorganic chemistry, *Gmelin* is preferred by this author. As compared to *Mellor*, it is better organized, more complete, more current, and easier to use.

Gmelin is organized according to what are designated as "Systems," numbered from 1 to 71. Figure 17 depicts the numbering system. As shown, any compound or combination of elements appears in the volume for the element having the highest system number.

Within each system, information included is subdivided into such categories as: general, preparation and manufacture, uses, analysis, production statistics, handling, and toxicity.

Although most of the text of *Gmelin* is German, the editors have taken several steps to facilitate use for chemists who lack some familiarity with that language. In some of the more recent volumes, one or more chapters are in English, and in a few cases (e.g. *Water Desalting*), the text of the volume is totally in English. Newer volumes provide English-language translations of tables of contents and of marginal captions. Also, Register (index) volumes have begun to appear in both German and English.

As noted, *Gmelin* is currently in its eighth edition. The "Main Work" (Hauptband) volumes, first published beginning in 1922, are augmented by the continuing publication of supplemental (Ergänzungsband) volumes for those elements whose original treatment was completed a long time ago.

Additionally, *Gmelin* has started to issue "Main Supplement" volumes. It is hoped that this approach will make it possible to contain the most recent information possible in a work of this type. In many of these volumes emphasis is on organometallics.

As an aid to using *Gmelin*, the chemist will find that the reverse of the title page for each volume contains the latest date through which literature for that system volume is evaluated. This means that the chemist can elect to begin searching of other sources, if needed, issued subsequent to that date and thereby save considerable time.

Gmelin provides key information in concise form. The original source of the data is referred to, as, in many cases, are abstracts of these sources.

One of the most significant features of *Gmelin* is inclusion of many valuable tables of numerical data, curves, and other graphic material, including diagrams of apparatus.

Considerable attention is paid to commercial practices in manufacture.

It is difficult to find fault with so thorough and competent a work as *Gmelin*. If there is any primary failing, it is lack of up-to-dateness. Even that is understandable in a publication so ambitious in scope, and major efforts are made to be as current as possible, especially in fields of high interest. For some of the material included, coverage is into the 1970s.

2. Mellor

As compared to *Gmelin*, *Mellor* (10-16) has the advantages of being completely in English (at least this is an advantage for chemists who read only English) and of a full (except for supplements) cumulative index, which *Gmelin* lacks at this writing.

The original volumes of *Mellor* were published over a time period beginning in 1922 and ending in 1937. In addition, the following supplements have been published:

- **a** *Volume 2, Supplement 1: The Halogens* (1956)
- **b** *Volume 2, Supplement 2: The Alkali Metals, Part 1* (1961)
- **c** *Volume 2, Supplement 3: The Alkali Metals, Part 2* (1963)
- **d** *Volume 8, Supplement 2: Nitrogen, Part 2* (1967)
- **e** *Volume 8, Supplement 3: Phosphorous* (1972)

3. Some Other Important Inorganic Chemistry Reference Works

In addition to such major treatises as *Gmelin* and *Mellor*, works intended to be of lesser scope but more recent coverage have been issued. In the field of inorganic chemistry, a good example is the four volume treatise by Bailar (10-17).

Inorganic Syntheses (10-18) contains checked synthetic procedures for inorganic compounds of "more than routine interest." New volumes issue every year or two.

KEEPING UP WITH AND IDENTIFYING BOOKS

Several important sources help the chemist keep up with the many new books, which appear with almost alarming frequency.

The most complete source is *CA*. Here, new books are cited in a special place in the sections in each current issue. The indexes to *CA* signify that an entry is to a book or encyclopedia by use of the designation "B."

Current journals provide lists of books and/or book reviews. For example *Chemistry and Industry*, *Chemistry in Britain*, *Chemical Engineering*, and *Journal of the American Chemical Society* contain excellent reviews. Extensive listings of new books appear in *Chemical and Engineering News*. The *Journal of Chemical Education* has a useful, comprehensive listing, categorized by broad subject, in its September issue each year.

The librarians who serve chemists have available a number of tools that can help chemists locate books in all fields of science and technology. For example, *Books In Print* (10-19) lists most English-language books; it is conveniently arranged by title, broad subject, and author.

11 PATENTS

In the United States granting of a patent indicates that the inventor has the right to *exclude others* from making, using or selling his invention in the United States, its territories and possessions, for 17 years after the patent is issued by the Patent Office.

A patent may be granted to the inventor of any *new* and *useful* process, machine, manufacture (includes all manufactured articles), or composition of matter (may include new chemical compounds or mixtures). For a more complete legal definition and explanation, the reader should consult publications (11-1) of the Patent Office or a patent attorney or agent registered to practice before that office.

Other countries have similar provisions. One major difference among the countries is duration of the monopoly granted, and another is speed of publication of the patent after filing the application. A third difference is ultimate validity, especially for some countries in which applications are not examined before initial publication (see p. 129).

WHY PATENTS ARE IMPORTANT

Chemists ought to display a keen interest in patents for two principal reasons:

1 Many chemists discover new products, processes, or uses that can be patented and can lead to substantial financial and other rewards for the inventor or his organization.

2 Patents are a rich and unique source of chemical information.

More specifically, the benefits of the patent as an information tool include these:

1 It permits the chemist to evaluate an idea rapidly and accurately.

2 It tells whether the idea belongs to anyone else.

3 It gives a feel for directions in which others are working, thus making it possible to evaluate one's own competitive situation.

4 It gives a trading position, for several inventors might have parts of an idea and might need each other to make an economical operation.

5 It is a good source of ideas, and especially component parts, to build a total system.

PATENT STRUCTURE

A brief outline of the parts of a typical United States patent will be helpful. These parts include:

1 The front page, which contains basic bibliographic information and an abstract.

2 The drawings (if any).

3 The so-called specification, which constitutes the main body of the patent (in terms of size) and contains a description of the invention.

4 The claims, which always appear at the end of the patent and delineate the scope of the monopoly granted under the law.

If the chemist reads the abstract and the first few paragraphs of the specification, he can identify the general purpose and nature of the invention. He should look for sections captioned "Background of the Invention" and/or "Summary of the Invention."

The specification usually contains in the first few paragraphs indication of the uniqueness (that which is new) and of the advantages of the invention over "prior art" (that which is already known.)

In most patents the specification includes "examples," which are so labeled. These examples are useful to the chemist because they give specific experimental details.

Claims express the legal essence of the patent and are of special interest to chemists in industrial organizations, particularly to patent attorneys, who are trained to interpret what is, and is not, covered by the patent from a legal standpoint. The first claims of most patents tend to be broad (generic); succeeding claims tend to become more and more specific. Almost all claims are, quite properly, highly legalistic in phrasing.

Official patent titles, particularly those of several years ago and many issued today, may be too general to be a meaningful guide to the technical content of a patent.

Good patent information tools, such as most of those mentioned in the chapter, try to provide the chemist adequate and reliable clues to patent content in several ways:

1 Patent titles are enriched with words that help make the context of the patent more understandable to the chemist. For example, titles can be enriched by adding words of substance, e.g., "Compound XYZ, *A Solvent, Manufacture by Process ABC Using Catalyst C.*" The italicized words represent hypothetical enrichment of a hypothetical title.

2 The content of patents is indicated by indexing of key features, especially examples and novel subject matter covered by the invention.

THE ROLE OF PATENTS IN IDEA GENERATION AND CREATIVITY

Patents are invaluable in helping guide chemists in evaluation and development of ideas. Although the patent system is available to chemists looking for information, it is frequently not used to the extent that it should be. Effective use of patents can lead to more productive research and development in chemistry. Such use often requires the cooperative efforts of the research chemist, the patent attorney, and the literature chemist or information specialist.

Since the beginning of time man has had an urge to create. By "create" this author means to carry a new idea to the public. This creation may take the form of a book, a painting, a musical arrangement, a floral design, a structure such as a house or bridge, or in the case of chemistry, a new chemical compound, process, or use.

Those who create things usually want to be identified with them. All men and women recognize that progress is made by these creative people. The government has seen the need to develop systems to encourage them to create. The three most important ways used by the government to encourage creativity are:

1 The copyright for writing and music
2 Trademarks
3 Patents.

In the patent system the creative person describes his concept, files the concept with government patent offices in one or more countries, has it examined by others (officially designated government examiners in the United States and many other countries) to determine if it is really new, inventive, and useful, and if it is, to grant him the right to exclude others from using the concept or invention for 17 years in the case of patents granted by the United States Patent Office, as noted at the beginning of this chapter.

The purpose of granting patents is to *encourage* people to create. It is anticipated by the government that these concepts will be *used*.

The patent system was developed to tell chemists and

others what has been going on and what areas of technology have been staked out by others. So the patent system was, and is, not only a *legal* tool or system, but an *information* tool.

Not to use the patent system is a mistake, because if an idea is old, the sooner the chemist knows it the better. Even if the idea is old and patented, the chemist might still be able to use it. For example, the first person to conceive a specific chemical idea could have been 20 years ahead of his time. If this is so, then present-day chemists could have the concept "free." Also, it may be that the basic concept was conceived, but something was missing to make it useable. This may permit a modern-day chemist to modify an older invention to make it useful. This is building on the technical foundation of the past and is what the creators of the patent system wanted to happen.

PATENTS AS INFORMATION TOOLS

How can the chemist use the patent system as an information tool? Searching the literature, reading, and thinking are cheaper and faster than experimenting. So, a chemist's approach to a problem could be as follows:

1. Think and guess what he would like to accomplish.
2. Make a quick survey of patent and other scientific literature.
3. Study.
4. Think and guess again.
5. Make a more detailed study of patent and other scientific literature.
6. Read, study, think.
7. Experiment.

To do the literature searching noted above, the laboratory chemist needs expert help—people who know their way around the patent system and other parts of the scientific literature. The patent system has become too complex for most laboratory chemists to "go it alone." That is why it is important for the laboratory chemist to confer with the information chemist, tell that person what is being thought and guessed about, and ask him for help in pulling together the works of others for study.

In this type of interactive relationship, the laboratory person tells the information person the type of search needed and also suggests how much time is worth spending on the search. Time to be spent is based on such factors as potential profit or other impact and priority considerations.

Next, the laboratory chemist studies the results of the search, does some searching himself, and makes decisions based on these studies.

His decision could be one of these:

1 Dig a little deeper.
2 Modify the original idea—shift to the left or right.
3 It's "old stuff"—forget it.
4 It's "old stuff," but by adding A or B it can be used.

The benefits of the above approach include learning, using appropriately the skills of others, and determining as quickly and economically as possible if an idea is good and novel.

The next, or perhaps concurrent, step could be to work in close cooperation with a patent attorney or agent toward the ultimate filing of a patent application if that is perceived as a goal.

PATENTS VERSUS JOURNALS AND BOOKS; QUICK-ISSUE PATENTS

Patents are a major information source about new chemical products, processes, and uses. The 24 leading patent-issuing countries issue approximately 12,000 patents each week according to some estimates.

Although the journal or book is more familiar to some chemists, this author believes that patents are now on a par with books and journals in value of information provided.

Some people do not look on patents as a source of reliable information. It has, however, been the experience of many chemists that it is as easy to replicate an example in a patent as it is to duplicate an experimental description in a scientific journal. In both cases, the inventor and the scientist usually report the highest yield obtained, but in both cases this yield was not always obtained in exactly the way the process was described. Nevertheless, with a little ingenuity and a knowledge of the field, the chemist can easily duplicate the majority of examples in United States patents. Examples in foreign patents may be a little harder to duplicate, but not if the chemist is skilled in reading between the lines.

Because chemists do not usually pubish anything proprietary in journals and books, at least until there is fully adequate patent coverage, patents are usually good clues to what competition is thinking.

Also, it is believed that much, but by no means all, of what appears in issued patents cannot be found published in other forms.

More importantly, many believe that patent literature has *replaced* conventional trade and scientific journals as the most *up-to-date* source of information on technological progress in numerous fields. This is because many countries have revised their laws so that patent applications are published without examination a few months after filing. Such countries are called "quick-issue" countries, and most are listed on p. 136.

The examination system used in the United States requires patent examiners to evaluate all applications on the basis of such criteria as unobviousness, novelty, and usefulness. These criteria are more specifically defined by law.

The examination process, to be done well, takes time. In

the U.S., the current *average*, from filing of the patent application to issuance of the patent or abandonment, is about 19 months, with a projected goal of 18 months. If the current rate can be maintained and the target ultimately achieved, this would put the accessibility of an *average* issued U.S. patent within the time range of patent applications published by some so-called quick-issue countries. The word *average* is emphasized because if the state-of-the-art in a field is "close," issue of the U.S. patent could take considerably longer than 18–19 months, whereas in other fields, issue could be much faster. (For example, inventions relating to energy or to environment could issue in six–eight months after filing, if expedited examination is requested at the time of filing.)

By screening quick-issue patents from the more rapidly publishing countries, chemists can gain advance knowledge of inventions whose counterparts (equivalents) may be issued later in countries that publish more slowly.

The chemist in North America and in the UK can better understand and appreciate the significance of quick-issue patents—particularly those from West Germany and Japan—if he is aware of recent studies at the U.S. Patent Office and elsewhere which show significant patent activity in the U.S. by inventors in these two countries, based on 1976 data.

Because the current "track record" (or performance) of countries such as West Germany and Japan is impressive, any chemist who wants to keep up with the latest technology is advised to follow the information sources covering these quick-issue countries and, as time permits, those for other major quick-issue countries. Close watch of foreign patent activity is particularly important in industry, but can be of equal value in academic and other organizations.

Despite all that has been said above about some of the advantages of quick-issue foreign (non-United States) patents, issued United States patents are frequently of great importance and deserve the careful attention of chemists in the United States and in other countries. Significant foreign inventions are usually patented in the United States if possible, because it is such an important market.

Patents are unique sources of chemical information in that they are primarily legal documents. Some chemists find patents more difficult to read and understand than books and journal articles because patents are usually written in legal phraseology by patent attorneys or agents who act for inventors. But this should not deter the chemist from taking full advantage of the unique and rapid access to the enormously broad range of chemical information which patents provide. One key is to read patents in the manner suggested on p. 124.

Similarly, patents should not deter chemists from using journals, books, and other forms of chemical information. All forms are important, and all have certain advantages which vary with the situation and intended use.

OFFICIAL GOVERNMENT SOURCES OF PATENT INFORMATION

An important way for chemists to learn about new patents is reading the patent gazettes usually published weekly by patent offices of most major industrialized nations. Here a chemist will find the principal claims of new patents issued each week.

A chemist can use the gazettes to keep up-to-date with patents; this is usually the quickest way for any single country. But this can be difficult for a variety of reasons. One is that key patents may turn up in material issued by unexpected countries and in unfamiliar languages.

Furthermore, many chemists find that information given in the gazettes is less useful than the patent abstracts which appear in such services as *Derwent* and *Chemical Abstracts* described elsewhere in this chapter. The gazettes, however, provide some information useful to laboratory chemists, for example, in helping decide whether to obtain a copy of the full patent or to try to locate an abstract that would provide more information.

OBTAINING INFORMATION ABOUT UNITED STATES PATENTS

The *Official Gazette*, compiled by the United States Patent Office, provides a listing of claims (usually only the main claim) for all United States patents issued each week. Copies of the *Gazette* can be found in many university and public libraries and in almost all industrial research organizations.

Patents listed in the *Gazette* are arranged in three major categories:

1 General and mechanical
2 Chemical
3 Electrical

Note that there may be some overlap among these categories. A patent of chemical interest may be found in one of the other sections.

Within each category, patent claims are grouped in numerical sequence according to a classification system established and maintained by the patent office.

To use the *Gazette* effectively, the chemist must identify pertinent classes and subclasses that reflect the interests of the chemist or his organization. This identification can best be achieved with the aid of a patent attorney or by direct communication with the United States Patent Office, since the classification system is complex and is revised at intervals.

If a specific inventor or organization is of interest, access by this key is relatively easy. It involves scanning the indexes to the *Gazette* which appear weekly and are cumulated annually.

The annual indexes to the *Gazette*, however, have been appearing many months after completion of the calendar year. Beginning in 1976 a commercial service started publishing quarterly and annual indexes to help bridge the gap (11-2).

Furthermore, the annual index to the *Offical Gazette* does not contain a subject index, although all United States patents issued that year are classified by class and subclass. This feature is useful to the patent attorney or agent. It can be a difficult means of access for the working chemist because of the complexities of the classification, as noted above, but use becomes much easier if the chemist studies the classification and knows his field.

When looking for organizations in patent indexes the searcher must consider corporate name changes. Companies merge, acquire partially or totally owned subsidiaries, disinvest or "spin off," or change names to reflect new activities and policy changes.

For example, a historical search over a period of years for patents issued to one company might include such variations as:

1 Mathieson Alkali Works
2 Mathieson Chemical Corp.
3 Olin Corp.
4 Olin Industries, Inc.
5 Olin Mathieson Chemical Corp.

Another way to search for and identify United States patents is to visit the Public Search Room of the Patent Office located at Crystal Plaza, 2021 Jefferson Davis Hwy., Arlington, VA.

This is the only place where all United States patents are arranged by subject matter according to a *Manual of Classification* (11-3) which helps the user decide which of over 300 main categories or "classes" to search. These classes are further broken down into more than 90,000 subclasses.

The searcher removes the selected subclass bundles of patents from their open-stack arrangement, takes them to a desk, and inspects or reads them in their chronological order.

If a patent discloses subject matter classified in two or more subclasses, a copy is placed in each subclass. To distinguish these copies from each other, the copy placed in the subclass selected as the principal basis of classification is called an original, and all others are called cross-references. Both original and cross-reference patents are contained in classification bundles in the search room.

Note that almost all United States patent applications being examined for possible issue are *not* available to the public. (Exceptions to this rule include all patent applications developed by the United States government which meet certain criteria; these applications are published by the U.S. National Technical Information Service, Springfield, VA 22161.)

Although most United States patent applications are confidential, the quick-issue (quick publication) systems used in some other countries can often provide important clues to developments for which applications might be filed in the United States.

Unfortunately, the search room is geographically inconvenient for regular use by most chemists, except those who live in the Washington, DC area, or whose business brings them to that location. All are, however, always welcome to use this facility.

For some years now, there has been talk in government and private circles about establishing one or more counterparts of the existing search room in other locations. This concept is still "alive," but no decisions have been made on whether the idea will be implemented. Chemists would do well to follow any discussions on the concept, because the result could be more ready access to categorized patent files for many.

Although the search room is available to all, this author believes it is designed primarily for use by patent attorneys, patent agents, and other trained patent searchers. Laboratory chemists and other inventors desiring to use the search room for the first time should seek assistance from the staff of the room, or, ideally be accompanied and guided by a patent attorney or agent from his own organization who knows all patent office facilities and how to use them efficiently. Chemists and other search room users will make good use of a patent office booklet on how to use the search room and associated facilities. This booklet is scheduled to be issued in 1978.

In addition to the patents in the public search room, unofficial subclasses and foreign patents and literature can be found in examiner search areas and may be searched with the permission of appropriate officials in these areas.

The scientific library of the patent office includes official journals of foreign (non-United States) patent offices and millions of foreign patents.

As a convenience for those who cannot visit its facilities in Arlington, the patent office will supply lists (patent numbers) of original or cross-reference patents contained in the subclasses comprising the "field of search." The chemist should seek the help of a patent attorney or of appropriate patent office experts in defining the various subclasses pertinent to his interests.

Copies of patents can then be inspected by chemists and others in a university or public library located within reasonable travel distance (for a list of these libraries see p. 42) which has a numerically arranged set of patents. Additional libraries are added to this list as indicated by needs.

Other significant sources of information about United States patents include Derwent, Chemical Abstracts Service, and IFI/Plenum Data Company; these are described in the sections that follow.

DERWENT PATENT INFORMATION SERVICES

From the previous sections the reader can appreciate the need for highly specialized services or centralized organizations to undertake prescreening processes and to make available classified or indexed informative lists of abstracts and patent numbers, suitably packaged for ready use.

Many chemists believe that the worldwide leader in coverage of chemical patent information is Derwent Publications, Ltd., a subsidiary of the Thomson Organization. Derwent is located at Rochdale House, 128 Theobalds Road, London, WC1X 8RP, England.

Much of Derwent's success and reputation in the patent information field is a direct result of the leadership of Mr. M. Hyams, their principal executive for many years.

1. Derwent Central Patents Index (CPI)

The Derwent *Central Patents Index (CPI)* is an abstracting and retrieval service dealing with chemical and chemically related patents.

Each week Derwent publishes booklets with English-language abstracts of current inventions arranged by country. The 12 "major" countries covered are Belgium, France, Japan, the Netherlands, South Africa, and West Germany, all of which publish patent applications just a few months after filing and are known as quick-issue countries; and the slower publishing countries, Canada, East Germany, the Soviet Union, Switzerland, the United Kingdom, and the United States. The new British Patents Law, which applies to applications filed after June 1, 1978, provides for early publication and will probably impact on availability of patent information from that country.

The 12 "minor" countries introduced since 1975 are the rapidly publishing (quick-issue) countries of Brazil, Denmark,

Finland, Norway, Portugal, and Sweden; and slower publishing Austria, Czechoslovakia, Hungary, Israel, Italy, and Rumania.

As examples of quick-issue access, consider the following:

a Belgian patent specifications are commonly "laid open to public inspection" in Brussels *without examination*. About half the documents are published within 2 to 3 months of filing; the rest within 6 months. This is usually before their counterparts anywhere else.

b West German patent applications are published weekly under the "new law" *without examination*, about 18 months after the earliest priority date—which for convention applications is normally about 6 months after the filing date. (Note that some chemists believe that published West German applications may be the most useful of the quick-issue patents, because copies can be obtained quickly and easily; if the technology and chemistry are important, an application will probably be filed in West Germany; the language is more familiar to chemists and translators than any other language with the exception of English; and, in addition to outstanding coverage by Derwent, West German patent applications are thoroughly indexed by *Chemical Abstracts*, *if* they are first issued in that country as noted on p. 63. *Examined* West German applications may be published up to 10 years later. This is a dramatic illustration of the importance of quick-issue (unexamined) patent applications.)

A week after the country-sequenced Derwent Altering Bulletins appear, the same abstracts arranged according to the Derwent subject classification system appear. The following week, a booklet giving more detailed, coded abstracts of "first-disclosure" and "basic" patents appears. (The concept of basics is discussed again in a few paragraphs.)

The chemist can retrieve this information through manual scanning of the printed product, including the abstracts proper, cumulative indexes by patent number, manual Derwent coded cards, punched Derwent coded cards, magnetic tapes that can be run for the chemist by his organization's computer center, and on-line access. Also available is a "Bureau Service" under which Derwent personnel will search their files on a current or retrospective basis.

Abstracts and complete specifications of all basic patents covered are available from Derwent in microfilm form.

In *CPI* patent documents are divided about equally between those that relate to entirely new inventions, or basics, and those for which corresponding patents have already been published in another country, or equivalents.

For the basics, Derwent editors prepare a meaningful descriptive title in English. Almost all patents are accompanied by English-language abstracts of considerable length; an example and/or drawing is frequently included.

Chemical patents currently account for about 45% of total Derwent coverage. They are selected for inclusion in one or more of the 12 sections:

A *Plasdoc* (covers polymers, fabrication, additives, uses, and specified monomers; began in 1966)

B *Farmdoc* (patents of pharmaceutical, veterinary, and related interest; began in 1963)

C *Agdoc* (compounds of agricultural and specified veterinary interest; began in 1970)

D *Food, Detergents* (includes disinfectants; began in 1970)

E *Chemdoc* (general organic and inorganic compounds and dyestuffs; began in 1970)

F *Textiles, Paper, Cellulose* (began in 1970)

G *Printing, Coating, Photographic* (began in 1970)

H *Petroleum* (began in 1970)

J *Chemical Engineering* (began in 1970)

K *Nucleonics, Explosives, Protection* (began in 1970)

L *Refractories, Glass, Ceramics* (began in 1970)

M *Metallurgy* (began in 1970)

All of these are available for on-line searching (see p. 82) to subscribers of the printed versions who are System Development Corporation (see p. 80) users and approved by Derwent. At a higher rate, nonsubscribers may also now access the on-line file, with the exception of special coding categories.

Note that *Derwent*, and probably other files, will be accessible on-line in Britain and the Continent through networks designed for that region, e.g., *InFoline* and *Euronet*.

To meet the needs of individuals having limited spheres of interest, a number of basic abstracts booklets covering quite narrow fields are issued as "profile booklets."

An outline of topics covered in all sections and booklets is given in brochures readily available from Derwent.

2. Derwent World Patents Index (WPI) and World Patents Abstracts (WPA)

The first announcement of a patent in any of the Derwent services is in World Patents Index (*WPI*)—a feature that makes this a unique and important publication. This is a collection of indexes giving details of all inventions covered in the 24 countries previously mentioned and classified according to patentee, subject matter, patent family, patent number, and priorities claimed.

WPI Gazettes are published in four editions—general, mechanical, electrical and chemical—based on the International Patent Classification (IPC). Abstracts are *not* included in *WPI*.

The weekly *WPI* gazettes are useful for watching for patents by patentee, for determining whenever a given invention of interest appears or reappears in another country, and for following by subject matter using the *IPC*.

World Patents Abstracts (WPA) is a collection of abstracts of inventions reported one week earlier in *WPI*. On a current basis, *WPA* takes the form of 10 different weekly publications, each devoted to a particular country, and six weekly journals, each dealing with a different area of technology on a multicountry basis.

The *WPA* journals, which are arranged by subject, cover only nonchemical subjects. Abstracts of chemical patents are provided by the *CPI*.

3. On-Line Access to Derwent WPI/CPI

The chemist can use Derwent's *WPI/CPI* files through on-line access. This feature was offered beginning in February 1976 using the facilities of System Development Corporation, Santa Monica, California. The current (1977) file consists of more than 1 million records and is growing at an estimated rate of 225,000 per year.

Chemists affiliated with organizations that subscribe to appropriate Derwent printed services can access the Derwent *WPI/CPI* data bank on-line through any or all of the following, either alone or in combinations:

Derwent accession number.

Derwent accession number year.

Derwent company code (this is the Derwent code for the company to which the inventor has assigned his patent; the company so designated is known as the *assignee*).

Derwent classes (these are broad categories that provide a degree of subject access).

International Patent Classification (IPC) numbers (these are more extensive categories than Derwent classes that provide additional access by subject category but not nearly in as much detail as the indexes of *Chemical Abstracts*).

Priority year and priority country (this information is primarily of value to patent attorneys and agents rather than chemists).

Patent number and patent country (this information can be useful in locating equivalents or patent families—thus if one has a

Japanese patent number and wants to determine if there is an English-language equivalent, it can be done in this way).

Title terms (the official or legal title of the patent is always reworded by Derwent to be more meaningful; when the file was initially "loaded" into the computers, terms were generated automatically from these titles; however, it has now been found possible to edit all 410,000 title terms originally present to remove plural forms and many synonyms by compiling a thesaurus and authority list so that terms on file will be more meaningful and searching will be much more certain; full availability of this feature is scheduled for 1978).

Derwent manual and punched codes (these codes help provide access by such means as structure and principal uses; use of these codes requires study and practice to attain proficiency).

4. Efficient Use of Derwent

Optimum use of Derwent—searching it efficiently both on a retrospective and current basis—takes special knowledge, best acquired by starting with attendance at training classes which Derwent offers at intervals, followed by ongoing experience. Also, up-to-date user manuals are available.

From a current awareness standpoint, however, it takes no special skill for a chemist to scan and read any of the printed Derwent bulletins. Using these bulletins, a chemist can keep up with a broad range of new developments in chemistry and chemical technology on virtually a worldwide basis. Although no chemist has the time or need to read all Derwent bulletins, he can select those of most interest to him, and within those bulletins, limit himself to categories of principal interest.

5. Advantages and Disadvantages of Derwent

The principal advantages of Derwent services are these:

a Coverage is comprehensive and virtually worldwide. Derwent editors interpret the definition of what constitutes a chemical patent broadly, so that more material is likely to be found in Derwent than any other comparable service.

b Abstracts are given for both equivalents and basics.

c Abstracts are lengthy and detailed, often including an example and/or a drawing. All abstracts are in English.

d Speed of coverage is excellent.

e Indexes to assignees are helpful in locating patents assigned to a given organization.

f Weekly and cumulative indexes that show related families of patents (or equivalents) are thorough and especially useful.

g The Derwent Patents Copy Service saves time and minimizes or eliminates the red tape and frustration usually associated with obtaining patent copies.

h Abstracts contain patent-filing details and other related legalities not ordinarily found in any other comparable service. This is of special value to patent-oriented chemists and patent attorneys.

Some disadvantages of Derwent include:

a There are no conventional printed substance, subject, or formula indexes such as those of *Chemical Abstracts*. Instead, the special Derwent codes and classification systems must be used, and title term access is available on-line.

b If the chemist's organization subscribes to *all* parts of Derwent, the service can be an expensive addition. This may mean that only organizations with excellent budgets can afford to get *all* parts of this important service. However, *CPI* subscribers who take at least 4 of the 22 subscription "units" have complete access to all sections on-line.

c Time coverage of the service is relatively limited; in the case of general chemistry, the beginning of truly comprehensive coverage goes back only to about 1970.

d Many chemists find overall use of Derwent and its components difficult, despite the availability of printed manuals and training classes on appropriate use. This difficulty can be overcome by careful ongoing study by the chemist or information specialist who intends to become a regular user of Derwent and to use it to maximum efficiency.

e There are more typographical errors (such as misspellings) than one would expect to find in a publication of this caliber. This situation will probably be largely corrected shortly.

f Abstracts are not available on-line. But citations are given to the printed versions of Derwent publications which do contain abstracts.

COVERAGE OF PATENTS BY *CHEMICAL ABSTRACTS* (*CA*)

CA is a major source of chemical patent information. Some aspects of treatment of patents by *CA* are touched on in Chapter 6, and elsewhere in this chapter. It is estimated that patents constitute about 30% of all documents cited by *CA*; the largest fraction of these cites is to unexamined Japanese patent applications, which began an explosive growth trend in 1972.

CA claims to cover all patents of chemical and chemical engineering interest for certain countries, as shown in Figure 18. For other countries, as shown in the figure, there is coverage only for chemical and chemical engineering patents issued to *resident* individuals or organizations in those countries (i.e. for *national* patents only). Beginning in 1978 (Volume 88), *CA* will extend its coverage of both *national* and *nonnational* patents to include Israel, Rumania, Switzerland, and the USSR.

Unexamined West German patent applications are designated as *Offenlegungsschriften*. Examined German applications, which appear later, are designated as *Auslegeschriften*.

PATENT COVERAGE

The following table summarizes *CA* patent coverage. For a more complete description of *CA* patent coverage for a particular volume or for patent coverage prior to Volume 74, the reader is referred to the introductory material at the beginning of Issue 1 of each Volume. For details on how to obtain patent documents, see the introduction in *CA* Issue 1 of the current volume.

Patent Coverage Summary

Type of Patent	Abbreviation	Year	1971		1972		1973		1974		1975		1976		1977		1978	
		Vol.	74	75	76	77	78	79	80	81	82	83	84	85	86	87	88	89
Australian	Austl		R	R	R	R	R	R	R	R	R	R	R	R	R	R	R	
Austrian	Aust		R	R	R	R	R	R	R	R	A	A	A	A	A	A	A	
Belgian	Belg		R	R	R	R	R	R	R	R	A	A	A	A	A	A	A	
Brazilian	Br Pl													A	A	A	A	
British	Brit		A	A	A	A	A	A	A	A	A	A	A	A	A	A	A	
British Amended B	Brit B													A	A	A	A	
Canadian	Can		R	R	R	R	R	R	R	R	A	A	A	A	A	A	A	
Czechoslovakian	Czech		R	R	R	R	R	R	R	R	R	R	R	R	R	R	R	
Danish	Dan		R	R	R	R	R	R	R	R	R	R	R	R	R	R	R	
Finnish	Finn		R	R	R	R	R	R	R	R	R	R	R	R	R	R	R	
French	Fr		A	A	A	A	A	A	A	A	A	A	A	A	A	A	A	
French Addition	Fr (5 digit number)		A	A	A	A	A	A	D	D	D	D	D	D	D	D	D	

French Medicinal	Fr M	A	A	A	A	A	A	D	D	D	D	D	D	D	D	D
French Addition to Medicinal	Fr CAM	A	A	A	A	A	A	D	D	D	D	D	D	D	D	D
German (East)	Ger E	R	R	R	R	R	R	R	R	R	R	R	R	R	R	R
German	Ger	A	A	A	A	A	A	A	A	A	A	A	A	A	A	A
Hungarian	Hung	R	R	R	R	R	R	R	R	R	R	R	R	R	R	R
Indian	Ind	R	R	R	R	R	R	R	R	R	R	R	R	R	R	R
Israeli	Isr	R	R	R	R	R	R	R	R	R	R	R	R	R	R	A
Italian	Ital	R	R	R	R	R	R	R	R	R	R	R	C	C	C	C
Japanese	Jpn	R	R	R	R	R	R	R	R	R	R	R	A	A	A	A
Japanese Kokai	Jpn K						R	R	R	A	A	A	A	A	A	A
Netherlands	Neth	R	R	R	R	R	R	R	R	A	A	A	A	A	A	A
Norwegian	Norw	R	R	R	R	R	R	R	R	R	R	R	R	R	R	R
Polish	Pol	R	R	R	R	R	R	R	R	R	R	R	R	R	R	R
Romanian	Rom	R	R	R	R	R	R	R	R	R	R	R	R	R	R	A
South African	S Afr	A	A	A	A	A	A	A	A	A	A	A	A	A	A	A
Spanish	Span	R	R	R	R	R	R	R	R	R	R	R	R	R	R	R
Swedish	Swed	R	R	R	R	R	R	R	R	R	R	R	R	R	R	R
Swiss	Swiss	R	R	R	R	R	R	R	R	R	R	R	R	R	R	R
U.S.S.R.	USSR	R	R	R	R	R	R	R	R	R	R	R	R	R	R	A
United States	US	A	A	A	A	A	A	A	A	A	A	A	A	A	A	A
U.S. Defensive Publications	US T	A	A	A	A	A	A	A	A	A	A	A	A	A	A	A
U.S. Patent Applications	US *							A	A	A	A	A	A	A	A	A
U.S. Published Patent Application	US B									A	A	A	A	D	D	D
U.S. Reissue	US R	A	A	A	A	A	A	A	A	A	A	A	A	A	A	A

A all patents of chemical or chemical engineering interest.
R chemical and chemical engineering patents issued only to individuals or organizations resident in the granting country.
D documents discontinued by issuing government.
C coverage discontinued.

Figure 18 **Chemical Abstracts** ***patent coverage.***

The examined German patent applications bear the same numbers as the final granted patents, which are designated as *Patentschriften*. *CA* indicates the unexamined patent by the abbreviation "*Offen.*" No differentiation is made between the examined application and the final granted patent in *CA*.

CA indicates the unexamined Japanese patent application as "Japan. Kokai." The examined Japanese patent application is designated as "Japan."

CA handling of West German and Japanese patent information is especially significant. For both countries *CA* provides extensive coverage and thorough indexing of unexamined patent applications. The significance of these two countries is based on their technological strength and on their quick issue policies noted earlier in the chapter.

Compared to some other patent information sources and tools, *CA* is characterized by such features as the following:

1 *CA* is probably more widely available and used in all types of organizations and libraries than any other comparable tool.

2 Indexing and abstracting of patents is outstanding.

3 For *CA* to cover a compound reported in a patent, there must be sufficient scientific or technical evidence; so-called "paper chemistry" is not enough.

4 More emphasis is placed on the chemical aspects than the legal aspects, with emphasis on details found in examples.

5 Although speed of coverage is good, the primary thrust appears to be the aspects noted above and accuracy.

IFI/PLENUM DATA COMPANY PATENT SERVICES

For the chemist interested in United States patents, and their equivalents in other countries, a key source of information is the IFI/Plenum Data Company, 2001 Jefferson Davis Highway, Arlington, VA 22202. A variety of services is available.

1. *IFI Comprehensive Database to United States Chemical Patents*

This service now contains approximately 400,000 chemical and chemically related patents dating back from 1950 to within a few months of the most recently issued patents. Access points (searchable) include: IFI accession number, United States patent number, assignee, Patent Office classes and subclasses, chemical compounds, chemical fragments, and so-called general terms. The file is distinguished by such features as depth of indexing (the current average is reportedly 50 descriptors per patent), relative ease of structure searching, and broad span of time covered. There are several other benefits which IFI officials can point out to potential users. The complete data base is best searched by computer.

Access is possible in several ways:

a IFI operates a service bureau that will conduct searches at a fee. Contact can be made by letter or wire, but a telephone call is also desirable, because search scope and strategy can best be developed this way.

b In some large research organizations the chemist can have access through his organization's computer center which can conduct searches as needed. Because this data base is expensive, relatively few organizations have this valuable file in-house at this writing.

2. *Uniterm Index to United States Chemical Patents*

This is a printed version of the *Comprehensive Database* described above. One advantage is an arrangement for manual (visual) searching—a computer is not used. Also, the cost is considerably less than the *Comprehensive Database*; this is a tool affordable by more organizations. One disadvantage is that it includes only about one-half the depth of indexing of the previously mentioned IFI tool (although even at that relatively lower level, the indexing is respectable). Another disadvantage is tedious use, involving frequent visual comparison of long columns of numbers; this difficulty can be eliminated by assigning the comparison chore to a clerk once the chemist has selected key words to be searched.

3. *The IFI Assignee Index to United States Patents*

This service provides access to United States patents alphabetically either by inventor or assignee from 1975 forward. Within company or organization (assignee), patents are listed in ascending order by class and subclass. Along with the patent number, the data given include the expanded title and a letter denoting type of claim [e.g., composition (C), process (P), and machine and device or article of manufacture (M)].

As compared to the index to the *Official Gazette*, this service is more rapid: it is issued quarterly and cumulated annually. It provides immediate (one-step) access to patent title and number—a process that requires two steps in the *Gazette* index. There is a printed version, and on-line access is available through the *Claims* service mentioned below.

4. *Claims*

On-line access to parts of the IFI system is available through Lockheed Information Systems in a file known as *Claims*. This covers the same patents listed in the *Uniterm Index* and the *Comprehensive Database*, but depth of indexing is much more shallow. Nevertheless, those features that are indexed and searchable, especially when coupled with coverage back through 1950, make this a handy tool. Searchable features include: title (official title through 1971, but enriched or expanded thereafter), assignee, class and subclass, *CA* reference when available, foreign patent numbers (equivalents) for certain countries, United States patent number, and other information. The ability to compile rapidly a list of patents to a specified organization is one example of uses to which chemists can put this file.

A related file, known as *Claims Class*, is an on-line index to the *Manual of Classification* of the United States Patent Office.

5. *International Patent Alerting Service*

This newest of IFI services is covered on p. 152 of this chapter.

SOME ADVANTAGES AND DISADVANTAGES OF IFI/PLENUM SERVICES

Some advantages include these:

a Coverage extends back through 1950 for United States chemical and chemically related patents.

b The subject indexing is in depth—considerably more so, it is said, than for some comparable services.

c *CA* references are provided whenever possible.

d Beginning in 1972 the type of patent (composition; process; machine or device) is indicated.

e General and mechanical patents can be searched on-line (beginning with 1971). Many of these patents are important to chemists.

Some disadvantages include these:

a Emphasis (with the exception of the new International Patent Service) is on United States patents. There are, however, cross-references to equivalents from five countries, Belgium, France, West Germany, Great Britain, and the Netherlands.

b The text of abstracts is not given, although, as noted above, *CA* references are given when available. The printed *Uniterm Index* includes a reproduction of the claim as reported in the *Official Gazette*.

c From 1950 to 1971 the titles are mostly the official (legal) titles of patents. This practice was changed, however, in 1972, when key words began to be added to help make titles more meaningful. More than 80% of all titles are now modified.

d Some chemists find the key word indexing more difficult to use than that of a "conventional" index such as *CA*.

e Speed of issue of some of the publications is sometimes slower than users would like for a service of this importance.

OTHER SOURCES OF PATENT INFORMATION

Other sources of patent information include:

1 Specialized indexing and abstracting services, such as the *Abstract Bulletin* of the Institute of Paper Chemistry (see p. 71), useful for pinpointing patents in specific fields.

2 News magazines, such as *Chemical Week*, which in identifying new technology frequently note pertinent new patents associated with that technology.

3 Articles, especially review articles.

4 Encyclopedias and treatises, such as those mentioned in Chapter 10—good especially for the older patent literature which can still be useful.

5 Other books, especially those from Noyes Development Corporation, which bring together United States patent information in specialized fields.

6 Programs such as those mentioned in Chapter 15, such as the SRI Process Economics Program which does an excellent job of technoeconomic interpretation of important patents on the manufacture of many major commercial chemicals.

PATENT EQUIVALENTS OR FAMILIES

Many organizations file patent applications in more than one country, especially for inventions believed to have widespread commercial or other importance that transcends national boundaries. This kind of filing cannot be done lightly, because extensive costs can be involved.

Accordingly, if a chemist discovers that a patent application has been published or issued in a number of countries, he can be reasonably sure that the organization that filed the patent application may be practicing, or at least strongly considering practicing, what is taught in the patent on a commercial basis.

From the point of view of the chemist interested in finding and using chemical information that represents actual practice, what has just been said is important. It is equally important to note that identification of equivalents (or so-called patent families) can be useful in other ways. For example, in the section on translations we noted the importance of finding English-language equivalents of foreign-language patents, thereby obviating the need for a translation. Also, identification of a patent family indicates that the chemist probably does not need to obtain copies of all patents in that family; they are probably, although not necessarily, similar in technical content.

How can the chemist most effectively achieve identification of patent equivalents or patent families? This is best done through use of one of several patent concordances.

One good patent concordance system is that of Derwent. It is issued weekly and cumulated frequently. A disadvantage of the Derwent patent concordance system is that it does not go back far enough in time—only until about 1970 in many cases.

The patent concordance to *CA* is issued weekly, but it is cumulated only semiannually. It goes back to 1962 and is probably more readily available to most chemists than any other concordance.

The most complete patent equivalent coverage for a specified time period is widely believed to be that offered by the International Patent Documentation Center (INPADOC) located in Vienna, Austria. Chemists in North America can conveniently access INPADOC files through the International Patent Alerting Service of IFI/Plenum Data Company. For 21 countries, coverage goes back as far as 1968. Beginning in 1973, the file was extended to include a total of 43 countries. Limitations of the file include its comparatively recent time span. Also, it is the only patent concordance of those mentioned not searchable on-line. IFI's other capabilities in searching for equivalents were mentioned earlier.

FUTURE OUTLOOK

A number of developments, some relatively close at hand, others on the distant horizon, need to be watched in the patent arena.

One is a potential further increase in the number of countries that opt for quick-issue systems. It is this author's guess that this is not likely to happen on a significant scale because of the number of countries that are already quick issue and because of some other developments as noted below.

For example, we probably can expect increasing use of computer-based patent searching systems by the larger patent offices. In the United States Patent Office, a computer-controlled microform search system is being developed. Important benefits both to the office and the public can be expected *if* a full-scale system can be implemented. Several years of experience, availability of adequate funding, and other factors will determine the ultimate success—or failure—of this system.

At the international level, several developments may have some impact on chemical patent information. One is the International Patent Cooperation Treaty. This is scheduled to go into effect on January 24, 1978. Filing of applications is scheduled to begin in mid-1978. On the basis of one international application, United States patent applicants can designate as many treaty-member countries as they wish at the time of filing. Such designation has the effect of lodging a national patent application in the patent offices of each of the countries designated. Patentability search reports prepared by the United States Patent Office will be forwarded to applicants and to the World Intellectual Property Organization in Geneva. Based on these reports, and other factors, inventors can decide whether to pursue the obtaining of patents in the various countries designated. They will have 20 months from the filing of an international patent application—if that was the first to be filed—to make their decisions. Use of this treaty will probably be mainly outside Europe.

The European Patent Convention came into effect on October 7, 1977. Uniform patent examination will be conducted at the European Patent Office in Munich which has been staffing for full operation by mid-1978. The output, following successful examination and granting, will be a "bundle" of national patents, each subject to laws and regulations of the European member country or countries for which a patent has been requested by the applicant. This "bundle" will be achieved on the basis of a single application.

Another development, which may be implemented in the 1980s, is the Community Patent Convention. This is closely linked to the nine-member European Economic Community. The proposal is that a single patent, valid for all member countries, will be granted, although each country could still grant its own patents as at present.

The effects of the plans noted will become more clear when the systems are fully in place and operational for a reasonable period of time. Some have speculated that the three agreements described will aid in the identification of equivalents, in the making available of more material in the English language, and in the assessment of claims. Chemists who use patents as information sources should try to follow at least the broad outlines of these international developments.

OTHER REMARKS ON PATENTS

- Any chemist who undertakes industrial or academic research today and has not first made a study of the state of the art (i.e., a patent search), in addition to looking at the other literature, is wasting his time. If he does not use patents as an information tool, he is not using what the government has made available to him. That has been a theme of this chapter and is worth repeating.

- Many chemists complain that there is too much "secrecy" in information and especially too much "secrecy" in chemical and other technical processes. That may be because they are not willing to study the patents and determine from them the current state of the art. Additionally, patents are an excellent stimulus for new ideas.

- One major goal of research, especially in industry, is to invent new products and patents that can withstand keen competition. Because competition does not stand still, patents and patent systems are a spur to ongoing research.

- Issuance of a patent does not necessarily mean that the invention will be commercialized; the patent may be primarily defensive (to exclude competition), or it could represent a potential interest that may never be exploited because of changes in policy, technology, or economics. Some patents are not commercialized immediately, but rather at a later date, perhaps a few years after issue of the patent, when conditions are more favorable for the technology or use.

- The United States Patent Office has a heavily used disclosure document program under which papers filed with that office in a prescribed manner may be used as evidence of dates of conception of inventions. Such papers are retained in confidence for 2 years and then destroyed unless referred to in a separate letter in a related patent application filed within 2 years. Persons desiring to use this program are advised to consult an attorney or agent registered to practice before the United States Patent Office.

▪ In 1969 the United States Patent Office instituted a defensive publication program whereby an individual or corporation may elect to publish an abstract of a patent application in the *Official Gazette* in lieu of examination by the United States Patent Office. On publication of the abstract, the applicant also agrees to open the complete application to inspection by the general public.

▪ This chapter has only skimmed the surface of the complexities of patents. For specific details, the reader should consult a patent attorney or agent—the author of this book is neither of those but rather has written from the viewpoint of chemical (not legal) information. Other good sources of information include, as previously mentioned, the United States Patent Office and its publications (11-4); Derwent user manuals (11-5); the book by Maynard (11-6); and the American Chemical Society's audio tape course on patents (11-7). These sources are merely examples of the voluminous literature and other tools about patents.

12 SAFETY AND RELATED TOPICS

The amount of information on safety, occupational hygiene, toxicity, environmental impact and control, and related aspects of chemistry and chemical engineering has been growing at a rate at least equal to that of work on newly reported compounds.

This kind of information is especially important before beginning work on any new project or on new or less-well-known chemicals, but it also has longer-range significance for all chemicals and all chemists. Pertinent information benefits not only persons directly connected with the investigation. It is equally important for colleagues in the same laboratory, pilot plant, or plant area and for all who may later use or otherwise come into contact with the chemicals.

Because the field is so large and of worldwide interest and action, this chapter emphasizes United States information sources. A recent paper by Churchley (12-1) is recommended as a good international review of safety information sources, including some material on regulations for the UK as well as other countries.

This chapter also emphasizes safety over environment information sources, although the two fields are closely related and may overlap.

The kinds of safety information with which the chemist or chemical engineer must be familiar includes one or more of the following examples:

1 Safe laboratory, pilot plant, and plant practices in general (applies to all chemists and chemical engineers).

2 Hazardous chemical reactions.

3 Flashpoint and other flammability or explosive characteristics. Also, smoke generation, if any.

4 Effects, if any, of specific chemicals on human beings.

5 Effects on test animals, especially if data on human beings is lacking.

6 Proper methods for pollution control, transportation, storage, and handling.

7 Effects on vegetation or aquatic life if discharged into the air or as liquid effluents.

8 Biodegradability.

9 Pertinent local, state, and especially federal standards and regulations issued by such agencies as the U.S. Environmental Protection Agency, Food and Drug Administration, Consumer Product Safety Commission, and Occupational Safety and Health Administration.

The field is dynamic, volatile. The standards, regulations, technological innovations, and toxicity/safety/environmental ramifications change rapidly. Because of this, chemists can expect presently available sources of safety and environmental information, as described in this chapter, to be supplemented or replaced continuously by new and improved sources. Organizations such as Chemical Abstracts Service and BioSciences Information Service, federal agencies such as the National Library of Medicine, and for-profit information industry organizations can be expected to take the lead in developing new information tools in this field of chemistry. Chemists or their librarians should do their best to maintain contact with these organizations on a regular basis and to keep up with the newer sources.

LOCATING PERTINENT SAFETY DATA

In industry, and to an increasing degree in universities, much of the needed data may already be available from departments in one's own organization. These departments are typically designated by such names as "Loss Prevention" or "Environmental Hygiene and Toxicology." Assistance of experts in these departments is essential in evaluating any potential hazards. Toxicological data, because of its biomedical nature, needs to be evaluated by a professional toxicologist or a physician.

In smaller, or less-well-organized laboratories, the bench chemist may need to make a personal investigation of pertinent safety data and should consider the following suggestions:

1 The manufacturer of the chemical, if it is commercially available, is a prime source of safety data and should be contacted as a first step. If trade literature is already on hand, the chemist must check with the manufacturer to ensure that this literature contains the latest information.

2 Since the published literature in the field is voluminous, consultation of basic reference books is a good next step. An exception to this sequence applies if an on-line, computer-based system, such as described later, is available to the chemist. If an on-line system is available, *this* is the preferred next step.

3 Federal, state, and local environmental officials may be able to provide advice and offer suggestions on safe handling.

4 The chemist can find some of the most recent published safety information, particularly on hazardous chemical reactions or laboratory procedures, in the "letters to the editor" and other sections of chemical news and other chemical magazines and journals. The chemist should scan these sections regularly and carefully, on a timely basis, if possible in cooperation with other chemists in the same laboratory, plant, or research-and-development team. Examples of good sources for this information include *Chemical*

and Engineering News, *Chemistry in Britain*, and *Journal of Chemical Education*. In addition, the *Journal of Chemical Education* features a series of columns on safety entitled "Safety in the Chemical Laboratory." These are currently edited by Norman V. Steere. The American Chemical Society's Division of Chemical Education has reprinted these articles in a series of three paperback books. (12-2) Major chemical accidents, such as may occur on a plant scale or during transportation, are reported quickly by the mass news media and then in more detail in the technical press. Some chemical organizations keep running tabs of these incidents, wherever they occur, to improve the overall safety of their *own* operations.

5 If safety data on the specific compound being worked with is not found, data on analogous or homologous compounds and other related structures can sometimes, but not always, provide useful guides to what can be expected. Such extrapolations are best made by a consulting toxicologist or other trained safety expert as appropriate.

6 In watching trends and developments in areas of safety, toxicity, and environmental concerns, the chemist needs to pay special attention to information and news from countries such as Japan, Sweden, Canada, and West Germany. These countries have shown special concern for the environment and are regarded as bellwether nations in this field. For example, the potential hazards of inappropriate use of mercury were first identified in Sweden and Japan.

7 Close to the leading edge of environmental developments are well-organized groups of informed citizens. Examples include:

a "Sierra Club"
530 Bush Street
San Francisco, CA 94108

b "Scientists Institute for Scientific Information"
560 Trinity Ave.,
St. Louis, MO 63130

which publishes the bi-monthly magazine *Environment*.

c "Center for Science in the Public Interest"

1757 S Street, N.W.,
Washington, DC 2009

which issues periodic newsletters.

These groups have been among the first in the United States to identify potential hazards. *Although not always correct in their assessment*, they can provide clues which may prove to be valid.

8 If no published information on hazardous properties of a chemical is found, this does not mean that no hazard exists. The chemist must assume that the chemical may be hazardous unless positive information is found showing that the chemical has been proven to be safe under specific conditions.

Questions which may help the chemist, assisted and advised by toxicologists, in evaluating the health risk (or lack of risk) that a chemical poses include these examples:

1 Will there be substantial or significant human exposure in:

a production
raw materials
intermediates
product
waste products

b distribution

c end use

d disposal

2 How does the chemical compare to existing products already in use which have similar properties and exposure characteristics?

3 How does it compare with such other existing products in physical properties that affect human exposure (vapor pressure, corrosiveness, etc.) and in toxicological properties?

4 Does it contain functional groups that are known carcinogens, mutagens, or teratogens?

5 Does a toxicologist have any reason to suspect that the product may pose an unusual risk in manufacture, use, or disposal that could not be minimized by adequate warnings or reasonable standard procedures? Experience with products in similar distribution patterns, end use, and disposal characteristics will also be valuable.

6 Will there be substantial exposure to the environment?

a Will it be produced in large quantities?

b How does it compare to existing products being distributed that have similar structure or toxicological effects and similar environmental exposure characteristics?

c Does it compare favorably in physical properties, molecular structure, persistence, solubility, or toxicity and degradation characteristics?

Because no single published source of safety data can be all inclusive, and for other reasons, it is a mistake to use only one secondary source. The source used could be biased, or the data could be taken out of context to mean something entirely different. The original source should be obtained and evaluated by an appropriately qualified safety expert such as a toxicologist (see also Chapter 3 on search strategy).

BOOKS AND RELATED INFORMATION SOURCES

Basic reference books on safety, toxic effects, and related aspects of chemicals are often the quickest and most efficient way to gather needed information. Several reference tools attempt to bring together in one place data on the safety aspects of a number of chemicals. Examples are cited in references (12-3) and a few are described below.

There are caveats: since any book can quickly become obsolete, particularly in safety-related matters, it is imperative that the chemist look at the most recent knowledge available and obtain as much pertinent detail as possible. Also, under today's conditions, no published compilation can cover the gamut of safety and related matters; none is complete.

The most widely known handbook on potentially hazardous chemicals is Sax's *Dangerous Properties of Industrial Materials* (12-4). Entries for over 13,000 materials are in the fourth (1975) edition.

Sax's book is excellent, but like all other books of its type, it is only a starting point. Not all materials or hazards are covered in the Sax book, and alternative information sources are sometimes more current or more complete. For example, the detail given in the MCA Safety Data Sheets (see p. 171) will not be found here. Also, Sax does not provide references to original literature or other sources for many of the materials listed.

Sax's book and other similar published works do, however, perform the important function of helping alert the chemist to some of the hazards—alerts that are then best followed up in detail in more specialized and up-to-the-minute sources, preferably in cooperation with safety or toxicity experts.

Bretherick's *Handbook of Reactive Chemical Hazards* (12-5) contains: stability data on specific compounds, data on possible violent interaction between two or more compounds, general data on a class of compounds or information on individual chemicals in a known hazardous group, and structures associated with explosive instability. References to sources of these data are given.

The most complete list of toxic effects of chemicals is the *Registry of Toxic Effects of Chemical Substances* (12-6) published annually by the U.S. National Institute for Occupational Safety and Health (NIOSH), Rockville, MD 20852. The 1976 edition contains toxicity information for more than 21,000 chemicals.

Beginning with the 1977 edition, the *Registry* will be

available in three formats: a microfiche version, which will contain the complete file updated quarterly; a two-volume printed book updated annually; and computer tape.

The file is searchable on-line through the National Library of Medicine. Additionally, full on-line access through the Fein-Marquart NIH/EPA Chemical Information System is now possible.

Note that on-line versions may contain some information not previously available in the printed volume. More importantly, the on-line version provides retrieval capabilities which permit the user to zero in on data that may be difficult to locate in other ways.

For most substances listed the following data are provided:

1 Chemical substance prime name (based on *CA* 8th Collective Index nomenclature).

2 Chemical Abstracts Service registry number.

3 Molecular weight.

4 Molecular formula.

5 Wiswesser Line Notation.

6 Synonyms.

7 Toxic dose data.

8 Cited references—sources of the toxic data.

9 Aquatic toxicity ratings.

10 Reviews.

11 Government standards and regulations.

12 NIOSH criteria document (for recommendation of a health standard at time of printing).

Before using the *Registry*, the chemist should read the introductory material carefully. From this he will learn that absence of a substance from the *Registry* does not imply that it is nontoxic and nonhazardous; the compilation is not complete. He will also learn that presence of a substance may not necessarily indicate that it must be avoided; here is an example where the chemist should call on the toxicologist for interpretation and guidance.

This author believes that involvement of the toxicologist in any use of the *Registry* is imperative; for the chemist to try to "go it alone" would be a mistake.

Now getting underway is the International Register of Potentially Toxic Chemicals (IRPTC). Work on this is being initiated under UN auspices in Geneva. The scope of this project will reportedly be broader than the NIOSH *Registry*. IRPTC will include data not only on mammalian toxicology but also on marine organisms and vegetation. Spills, persistence in the environment, principal uses, and general ecological effects will be included, as will cautions on transportation.

Two works of key importance to the environmental professional are *Standard Methods for the Examination of Water and Wastewater* (12-7) and *Methods of Air Sampling and Analysis* (12-8).

The testing procedures and standardized methods in these manuals help provide the consistent and reproducible results needed to comply with current requirements. Continued revision of both books can be expected as standards and technology change.

Both volumes are available from the American Public Health Association, 1015 18th St., N.W., Washington, DC 20036.

The proceedings of the Purdue Industrial Waste Conferences (12-9) are an invaluable collection of case histories on handling a wide variety of pollution abatement problems. These proceedings have been published annually for more than 30 years.

ON-LINE SOURCES

The computerized literature retrieval services of the National Library of Medicine (NLM) are available on-line through a system known as *Medlars*. Some significant components of this system are described below.

One of the most up-to-date and complete sources of toxicological and related information is the Toxline file of NLM. *Toxline* (Toxicology Information On-Line) is based on the National Library of Medicine's extensive collection of computerized toxicology information containing some 500,000 references to published human and animal toxicity studies, effects of environmental chemicals and pollutants, adverse drug reactions, and analytical methodology. This rapidly expanding data base contains full bibliographic citations, almost all with abstracts and/or indexing terms, and *CA* Service Registry Numbers from primary journals 1971 forward. Older information is in *Toxback*. *Toxline* information is derived from major secondary sources and special collections of material. The component subfiles currently providing *Toxline* material are:

1 Chemical Abstracts Service: *Chemical-Biological Activities* (CBAC), from 1965.

2 BioSciences Information Service: *Abstracts on Health Effects of Environmental Pollutants*, from 1972.

3 American Society of Hospital Pharmacists: *International Pharmaceutical Abstracts*, from 1970.

4 National Library of Medicine: *Toxicity Bibliography*, from 1968.

5 Environmental Protection Agency: *Pesticides Abstracts*, from 1966.

6 Hayes File on Pesticides 1940–1966 (citations only).

7 Environmental Mutagen Information Center, Oak Ridge National Laboratory, from 1968.

8 Toxic Materials Information Center File, Oak Ridge National Laboratory, 1971–1976.

9 Environmental Teratology Information Center File, Oak Ridge, from 1950.

10 Teratology File, Plenum, 1971–1974.

Note that although NLM headquarters staff are not usually available to perform toxicity searches for the general public, the Toxicology Information Response Center at Oak Ridge, a service arm of NLM, can perform a variety of searches, usually on a fee basis.

Other files available on-line through NLM include *Cancerlit* (deals with various aspects of cancer) and *Cancerproj* (includes on-going cancer research projects). Another file—probably the largest of the NLM files on-line—is *Medline* which contains hundreds of thousands of references to the biomedical literature.

NLM's *Chemline* is a file of several thousand names for chemical substances. Created in cooperation with Chemical Abstracts Service, it can help the chemist identify alternative names that may be used for the same compound. This is especially important when looking for toxicity and other safety data.

Additional NLM files are operational, but those mentioned illustrate the scope of what is now available.

NLM officials have plans for an on-line *Toxicology Data Bank*. This will include, for example: toxicological data (not just references), physical constants, shipping methods, manufacturing methods and plant locations, and extracts from so-called tertiary documents such as criteria documents, monographs, and handbooks. This is expected to become available in 1978. As compared to the *Registry of Toxic Effects*, only about 1000 chemicals will be included initially versus more than 20,000 in the *Registry*. The data, however, will be in considerably more depth, and plans call for careful selection and on-going review.

Chemists and others can use NLM files through 11 regional medical libraries in the United States and a number of non-United States centers. Additionally, there are many publicly and privately operated on-line search centers. More information can be obtained from the National Library of Medicine, 8600 Rockville Pike, Bethesda, MD 20014.

Examples of other tools containing safety information which can be searched on-line are mentioned elsewhere in this chapter. These include *CA*, *Biological Abstracts*, *Enviroline*, and *Pollution Abstracts*.

GOVERNMENT REGULATIONS AND RELATED

Most chemists and engineers are aware that there are many government regulations and standards relating to chemicals and the chemical industry. These government actions have, in many cases, significant impact on the public and on the way chemists and engineers do their work.

The interpretation and implementation of regulations have both legal and technical ramifications. Attorneys, toxicologists, engineers, and chemists who specialize in this field are needed to assure appropriate compliance. As in the case of patents, this is an area that is often best handled by specialists, although all chemists should have some basic familiarity with important developments.

General sources of information on government regulations and actions include:

1 Chemical news magazines. Also, major "national" newspapers, especially the *Wall Street Journal*. For the chemist or engineer to

whom regulatory affairs are not a full or part-time job, these sources should suffice.

2 The daily *Federal Register,* the official United States government announcement source for regulations of all types, including those pertaining to toxicity, other safety matters, and environmental affairs.

There are numerous specialized information tools and sources on the many and rapidly changing regulations and standards. One of the most outstanding of these is *Environment Reporter,* published weekly by the Bureau of National Affairs (BNA), 1231 25th St., Washington, DC 20037. The stated intent of this publication is to help subscribers:

1 Keep abreast of the latest developments on the environmental scene.

2 Be alert to what must be known about federal laws and regulations.

3 Be provided with major state-by-state air, water, solid waste, and land-use laws and regulations.

4 Be given full text of court decisions in important environmental cases.

5 Be supplied with analysis of specific environmental protection topics.

Even more specific to the chemical industry is BNA's *Chemical Regulation Reporter* which includes weekly reference information on important developments with special emphasis on the *Toxic Substances Control Act* of 1976.

Another good source is the Environment Information Center (EIC), 292 Madison Ave., New York, NY 10017. Products include the *Environment Regulation Handbook* (covers United States environmental laws and regulations) and state environmental laws and regulations in microform. On-line searching for environmental documents is possible through *Enviroline,* mentioned elsewhere in this chapter. There is further discussion of EIC products later.

Chemical Industry Notes (see p. 64) provides access to articles that report on government regulations.

The tools mentioned above, especially the compilations of the Environment Information Center and of the BNA, can also help the chemist make direct contact with appropriate high level government officials and technical experts by identifying the names of these individuals and their functions.

The most recent government agency organization charts are also important in keeping track of key personnel.

PROFESSIONAL SOCIETIES AND OTHER ASSOCIATIONS

Professional societies are active in promoting safety and providing safety-related information:

1 The American Chemical Society's Committee on Chemical Safety provides advice on safety matters and has published the booklet *Safety in Academic Chemistry Laboratories* (12-10). A Division of Chemical Health and Safety has just been formed.

2 The American Institute of Chemical Engineers has publications on safety, including *AIChE Pilot-Plant Safety Manual* (12-11), *Loss Prevention* (12-12), and *Ammonia Plant Safety* (12-13).

3 The Manufacturing Chemists Association (MCA), located in Washington, DC, provides a variety of services and tools to aid in safe handling of chemicals.

a A toll-free telephone number, 800-424-9300, provides immediate access to the MCA's Chemical Transportation Emergency Center (Chemtree). This center was established in 1971 to handle accidents with chemicals in transportation or warehousing. In addition to its original functions, Chemtree is also the communications contact for the Pesticide Safety Team Network and the Chlorine Institute's Emergency Plan. The service: (1) determines what chemicals are involved in an accident, (2) supplies preliminary advisory information retrieved from Chemtrec files, and (3) secures involvement of appropriate technical specialists.

b Safety Data Sheets issued by MCA cover about 100 of the most heavily used chemicals of commerce and other chemicals. The in-depth information provided is of value to chemists and chemical engineers in laboratory, pilot plant, and manufacturing operations. Users of this material should note that the most recent information is not always supplied; some data are several years old and need to be supplemented by more recent sources. As of mid-1976, many data sheets were in process of ongoing revision. Some have been temporarily withdrawn pending revision or clarification. A typical data sheet provides such information as: names; chemical and physical properties; training and job safety; health factors; fire hazards and fire fighting; instability and reactivity hazards and control; engineering control of hazards; storage, labeling, placarding, shipping, and handling; waste-disposal guidelines and spill control; and tank and equipment cleaning and repairs.

c Additionally, MCA has on record several hundred case histories of accidents and has published an index to these files. Recent accidents in plants or laboratories of chemical firms, along with preventive measures recommended by the company reporting the incident, are published several times a year in a special MCA bulletin, *Accident Case Histories*. Careful perusal of this material can help avoid future mishaps.

d The safety activities of the Manufacturing Chemists Association are carried out mainly by its Safety and Fire Protection Committee, formed in 1945. The committee wrote the comprehensive book *Guide for Safety in the Chemical Laboratory*, first issued in 1954. A revised edition was published in 1972 (12-14).

4 Like MCA, the National Safety Council, headquartered in Chicago, has a broad range of safety-related activities and publications. Many of these deal with safety in general or with other aspects of safety, but some, such as their *National Safety News*, have material of specific interest to chemists (12-3).

5 The chemist will find the National Fire Protection Association (NFPA), 470 Atlantic Ave., Boston, MA 02210 a valuable source of safety information. Of special interest is NFPA's *Manual of Hazardous Chemical Reactions* (12-15). This is a compilation of more than 3500 laboratory and plant incidents involving chemicals. Before the chemist or engineer embarks on a research-and-development or similar project, this NFPA publication is a key source to be examined. The NFPA publication *Hazardous Chemical Data* (12-16) contains basic safety data on several hundred chemicals. These data include: description (including odor), fire and explosion hazards (including flash point, flammable limits and ignition temperature), life hazards, personal protection, fire fighting, usual shipping containers, storage, and remarks (including representative standards and safety data sheets which may be applicable). Both of these publications are now (1977) in process of being updated.

In November, 1975, the National Fire Protection Association adopted a detailed fire protection standard for laboratories which use chemicals. This 54-page code (12-17) is one of the most comprehensive of its type. The product of years of work by the association's sectional committee of Chemistry Laboratories, the code deals with chemical handling and storage, compressed gases, fire prevention, ventilation, building construction, exit doors, and automatic fire-extinguishing systems. The new standard is expected to be used by state and municipal agencies in updating their fire protection codes.

6 The American Conference of Governmental Industrial Hygienists (ACGIH), Box 1937, Cincinnati, OH 45201 publishes threshold limit values (TLVs) for chemical substances in workroom air. ACGIH is not an official government agency, but its recommendations are highly regarded.

7 Published information on hazardous reactions is useful, but no matter how thorough, such materials cannot possibly include hazardous reactions that have never been tried. In response to this need, the American Society for Testing and Materials (ASTM) developed and are distributing a computer program (designated as CHETAH) developed by their committee on Hazard Potential of Chemicals. The program provides estimates of the maximum possible energy release for covalent compounds of carbon, hydrogen, oxygen, nitrogen and 18 other elements (see also p. 211).

8 The Chemical Society (London), like its United States counterpart, is active in the field of safety and related topics. One recent (1977) example is publication of the second edition of *Hazards in the Chemical Laboratory*, edited by G. D. Muir. Also, the Institution of Chemical Engineers has a long record of initiatives in safety. Their new guide *Flowsheeting for Safety* and their *Loss Prevention Bulletin* are two examples of pertinent activities.

9 A set of data sheets similar to that of MCA is being issued by the Chemical Industry Safety and Health Council of the Chemical Industries Association, London. Their publication *Major Hazards* (1972) identifies materials manufactured or used by industry that involve particular risk. They are now issuing *Codes of Practice* as part of a series relating to these materials and combining the experience and expertise of principal manufacturers as well as government personnel and major users.

10 The UK Fire Protection Association has published a *Guide to Safe Practice with Flammable Liquids.* This is intended for firms using small amounts of such liquids. The guide gives advice on design of premises, precautions in production and process areas, and staff training. The guide is published in the Fire Protection Association's journal, *Fire Prevention*, and is also available separately in reprint form to nonmembers as *FPA Guide to Safe Practice with Flammable Liquids.*

CHEMICAL ABSTRACTS AND OTHER ABSTRACTING AND INDEXING SERVICES

A vital source in this field of chemistry, as in all others, is *CA*. Some sections that can be scanned selectively on a current awareness basis include the following:

Section 1—Pharmacodynamics
Section 4—Toxicology

Section 50—Propellants and Explosives
Section 59—Air Pollution and Industrial Hygiene
Section 60—Sewage and Wastes
Section 71—Nuclear Technology

All 80 sections, however, may contain some material of interest.
Pertinent safety-related headings in the general subject index include:

Accidents
Combustibles
Disease, occupational
Electric shock
Explosibility
Explosion
Fire
Flammability
Health hazards
Health physics
Injury
Safety
Safety devices

These entries are helpful when looking for more generic information (such as safety in handling flammable organic liquids or the question of toxicity of surfactants to aquatic life). The same entries can also be helpful when looking for specifics (such as explosibility of o-nitrophenylsulfonyldiazomethane), although for specific substances the chemist should also look under the name of that substance in the *CA* chemical substance index.

Beginning in 1976, the keyword indexes to the weekly issues of *CA* were extended to cover safety more broadly.

However, this entry (the keyword "safety") by design does not include toxicology. Therefore, to get a complete picture of safe-handling aspects, the keyword index entries pertaining to toxicity must also be scanned. Because the keyword indexes are not yet (1977) complete for toxicity data, the best places to look in *CA* are the six-month volume and five-year collective indexes. This is true particularly for toxicity studies reported incidentally to synthesis of a new compound. *CA* editors are working with the National Library of Medicine and others to improve the situation.

In 1977 *CA* announced plans for a subset grouping of *CA* abstracts, *Chemical Hazards*, produced by computer search of the full *CA*, as part of the *CA Selects* series (see p. 23). Chemists and engineers should find this a convenient way of keeping informed about safety on a reasonably current basis, but it is too new for full evaluation.

Every two weeks *Chemical Hazards* presents abstracts on such subjects as:

- Health and safety of personnel working with or in the area of radiologically or chemically hazardous substances
- Hazardous properties of chemical substances and chemical reactions
- Effects of human exposure to and protection of humans from hazardous substances
- Toxicity of hazardous substances to humans
- Safety concerns in chemical laboratories and chemical industry

Excluded from coverage are toxicity studies done on laboratory animals.

In addition to *CA*, or the subset grouping noted above, the related publication *Chemical Industry Notes (CIN)* (see p. 64) also needs to be consulted for extracts of reports on specific incidents such as chemical plant explosions, fires, spills, leakages, and the like. If this material is "newsy," and reported in trade and industry publications or a few major "national" newspapers, it is more likely to be reported in *CIN* than in the full *CA*.

Current Abstracts of Chemistry (see p. 73) can prove helpful in alerting chemists to potential biological hazards of newly reported compounds.

Other products of the Institute for Scientific Information which can be helpful in following safety-related developments are their ASCATOPICS services (see p. 21).

An up-to-date and comprehensive source that emphasizes toxicity aspects is *HEEP (Abstracts on Health Effects of Environmental Pollutants)*, published monthly since 1972 by the BioSciences Information Service, Philadelphia, PA 19103.

Environmental chemicals and substances other than medicinals that affect human health are included, as well as general reviews and original papers reporting on potentially harmful effects on humans, lower vertebrates (used as indicators of the substances toxic to man), vertebrates and invertebrates in the food chain, and analytical methods for examining biological tissues or fluids. Monthly and annual indices facilitate use.

This is one of the sources included in the on-line *Toxline* service. For chemists who do not have ready access to *Toxline*, *HEEP* is a prime alternative for toxicity data.

A unique advantage of *HEEP* is that it covers not only possible adverse effects of chemicals on humans, but also, to some extent, the possible effects on wildlife, including fish. The importance of this point is readily apparent to any chemist attuned to contemporary environmental issues.

The full *Biological Abstracts*, issued by the same organization as *HEEP*, can contain some abstracts on biological activity that may not be reported elsewhere; coverage is from 1926 to the present. Manual and computer-based batch and on-line searches can be conducted, and a current awareness service is available. One of the principal merits of this file is its broad scope.

Environmental Health and Pollution Control, published by Excerpta Medica, P.O. Box 1126, Amsterdam, The Netherlands, is a monthly abstracting service that includes coverage selected from more than 15,000 scientific journals and periodicals. Chemists and engineers will find this particularly useful in covering developments in Europe, although the coverage is reportedly worldwide. Other related abstract bulletins in the Excerpta Medica series include *Microbiology, Public Health* and *Occupational Health*.

Environment Abstracts is another excellent source. It, too, is available in both printed form and on-line. The on-line version is *Enviroline* ® and can be searched by users of both Lockheed Information Systems and System Development Corporation.

The scope is broad. Categories covered include: air pollution; chemical and biological contamination; energy; environmental education; environmental design and urban ecology; food and drugs; general; international; land use and misuse; noise pollution; nonrenewable resources; oceans and estuaries; population planning and control; radiological contamination; renewable resources—terrestrial; renewable resources—water; solid waste; transportation; water pollution; weather modification and geophysical change; and wildlife. Copies of most of the documents included are available from the publisher, Environment Information Center, Inc.

Pollution Abstracts published by Data Courier, Inc., 620 South Fifth Street, Louisville, KY 40202, contains references to worldwide technical literature covering some 2500 journals, conference proceedings and papers, and monographs in the areas of air and water pollution, solid wastes, noise, pesticides, radiation, and general environmental quality. It is available both in conventional printed form and also on-line via Lockheed Information Systems and System Development Corporation.

There are at least two indexing and abstracting services that cover the "science of safety" defined broadly: *CIS Abstracts* and *Safety Science Abstracts*.

The International Labor Organization, Geneva, publishes the well-known *CIS Abstracts*, which many find useful for its international coverage of safety.

Safety Science Abstracts is a compilation of abstracts that have been selected, classified and indexed from the world's technical and scientific literature on the science of safety. Coverage includes events or phenomena that threaten mankind and his environment or the technology on which he depends. Equal emphasis is given to research and applications. Sources are from more than 8500 regularly scanned periodicals, as well as thousands of government reports, conference proceedings, books, dissertations, and patents. Each issue contains extensive cross references and complete indexes by subject, author, and acronym. A cumulative source index is published in issue 1 of each volume. It is published by Cambridge Scientific Abstracts, 6611 Kenilworth Ave., Riverdale, MD 20840, in association with the Center for Safety of New York University.

TOXICITY TO AQUATIC ORGANISMS

Potential toxicity of chemicals to fish and other aquatic organisms is of key environmental interest, especially if there is a chance that such organisms might come into contact with the chemical, or its by-products or waste streams, during manufacture or use. Government agencies and other sources have compiled much of the earlier and more recent published literature. For example, see reference 12-18. Recent data can also be obtained from some of the on-line sources and printed abstracting and indexing services mentioned earlier. Additionally, there are at least two somewhat more specialized sources:

- Aquatic pollution is one of the fields covered in *Aquatic Sciences and Fisheries Abstracts* published monthly by Information Retrieval, Ltd., London and Washington (1911 Jefferson Davis Highway, Arlington VA 22202.)
- *Selected Water Resources Abstracts* is an abstracting and indexing service published by the U.S. Department of Interior's Water Resources Scientific Information Center. This covers the subject of water quite broadly, including toxicity to aquatic life. It can be used in conventional (printed) form, and computerized access is possible through the facilities of several universities: Cornell University, North Carolina State University, University of Arizona, University of Wisconsin, and Virginia Polytechnic Institute.

Despite availability of these and other sources, there is a relative shortage of published data on the effects of chemicals on aquatic life. Mammalian toxicity data is more readily available.

Direct mail or phone contact with government experts on the forefront of this work can provide additional leads beyond what is in published sources. Examples of leaders in the field are scientists at the U.S. Environmental Research Laboratory, Duluth MN 55804.

BIODEGRADABILITY

The biodegradability of chemical substances is of considerable environmental interest. One of the most extensive known files on this topic is maintained by the Water Research Centre, Elder Way, Stevenage, Hertfordshire, SG1 1TH, England. This special service is known as INSTAB (Information Service on Toxicity and Biodegradability). The collection, which is critically appraised during its compilation, covers the effects of chemicals on biological sewage treatment processes, both aerobic and anaerobic. Also included are biological effects on freshwater invertebrates and fish. The service is mainly for members of the center, although nonmembers may pay to use the facilities. Among the many other activities of the center is publication of their broad-scoped weekly abstract bulletin *WRC Information*.

Other information on biodegradability will be found in such more general tools as *CA* and *Biological Abstracts* and in specialized environmental tools such as *Pollution Abstracts*.

WATER-QUALITY DATA

Water-quality data are important in appraisal and management of water sources, pollution surveillance and studies, monitoring of water quality criteria and standards, and development of energy resources.

One good source of water data is the National Water Data Exchange (NAWDEX). This is part of the U.S. Geological Survey, 421 National Center, Reston, VA 22902. NAWDEX was established in 1976 to assist users in identifying, locating, and acquiring needed data. It is a confederation of all types of water-oriented organizations. To cite just one example, coordination is being developed with the Environmental Protection Agency's STORET file which has waterways quality data from more than 200,000 collection points. The general public can access NAWDEX through its more than 50 local assistance centers located throughout the United States.

EXAMPLES OF KEY JOURNALS

There are many key journals in this field, but it may help the reader to mention a few examples:

- *Environmental Science and Technology* (American Chemical Society, Washington, DC)
- *Journal of the Water Pollution Control Federation* (Washington, DC); annual literature review issue is excellent
- *Journal of Hazardous Materials* (Elsevier Publishing Co., Amsterdam, The Netherlands); this new journal, of international scope, could prove especially important if it is continued with broadened coverage.

NEWSLETTERS

This field is characterized by a number of specialized newsletters. These may contain information that supplements and augments chemical news magazines and sources such as mentioned on p. 169. Emphasis in the newsletters is on rapid, concise reporting. One good example is *Air/Water Pollution Report*, published weekly by Business Publishers, Inc., P. O. Box 1067, Silver Springs, MD 20910.

OTHER

Much of the material in the next chapter has safety implications. See, in particular, the sections on evaluating and on determining or estimating properties.

13 LOCATING AND USING PHYSICAL PROPERTY AND RELATED DATA

Most chemists and engineers find it difficult to locate physical property and related data for new or little-studied chemicals or chemical systems. The same difficulty applies to location of critically evaluated data for almost any chemical—many chemists find these among the most difficult-to-locate kinds of chemical information. The location of evaluated, accurate thermodynamic data is especially challenging. If data are inaccurate or not fully understood, this can lead to erroneous conclusions or to syntheses or processes that do not work as expected. Proper use of accurate data is important in the design and engineering of pilot plants and plants that will safely produce chemicals of high quality at the desired yield. The laboratory chemist doing synthesis work finds good physical property data equally important in his work. Further, accurate data and correct interpretation of data are vital to major policy decisions at the federal level, such as on environment or energy matters.

The problem is complicated by the current policy of some journals not to publish extensive tables of data or full experimental details in an attempt to conserve printing and other costs. The trend is to make these data and details available in microfilm or other microform which sometimes needs to be purchased separately by the reader if he is interested. There are many scientists who consider this a false economy. They believe that information can be "lost" or misinterpreted by this practice, especially if the data microfilmed are not subject to normal editorial review.

Another factor is that some scientists are reluctant to evaluate the results of their work or that of others critically, or even—at a minimum—to provide sufficient information to permit others to recognize or evaluate the accuracy of their work.

For this and other reasons, some estimates are that for about 50% of data reported in scientific literature too little information has been reported for independent evaluation. This chapter addresses itself to some of the points outlined above.

SOURCES OF DATA

The sources and methods reviewed in this chapter can be arbitrarily grouped into several categories:

1 *Original Sources*—These are organizations or systems that generate or critically evaluate data. The National Standard Reference Data System is an example, specifically the system's data centers.

2 *Secondary and Tertiary Sources*—These are basically compilations of data from original sources and news about projects and activities. Examples include handbooks, journals, and books, including compilations and reports. Abstracting and indexing services, especially *CA*, are valuable for accessing pertinent data, and there are several special bulletins and newsletters in the field.

3 *Evaluation and Estimating*—In this field particularly, qualified chemists and engineers may need to do some of their own evaluation and perform some calculations using appropriate computer programs. Some organizations have designated specialists who concentrate on this activity.

RECORDING PROPERTY DATA

When any one property of a chemical is being searched for and is located, the chemist should make note of other properties (or at least the source of such properties). Experience has shown that this can save considerable future work, because several properties of a single chemical are often grouped together in one paper or other source.

Many chemists, in addition to recording physical property data in their laboratory notebooks, will also make use of a specially designed property data form. Such a form should be kept on file for each chemical and should have separate columns for recording the kinds of properties, values obtained, source, and date of source.

If a chemist is part of a team or group working together on the same project, all completed data sheets should be kept in a central place. This helps avoid repetitive efforts in which different individuals may look over and over again for the same values. A central repository for data is especially important in industry. For example, completed data sheets often become part of a process manual used by engineers to design and build pilot plants and plants.

Relatively new United States government regulations require use of a standard form (Material Safety Data Sheet) with emphasis on properties related to safe handling. This form can be adapted (expanded) to include a complete range of properties about which any organization wishes to maintain a verified, centralized record for all chemicals of interest.

NATIONAL STANDARD REFERENCE DATA SYSTEM

The National Standard Reference Data System (NSRDS) is one of the relatively newer key sources available to the chemist and engineer for obtaining the latest and best physical property and related data. NSRDS was established in 1963 as a means of coordinating, on a national scale, production and dissemination of critically evaluated reference data in the physical sciences.

Under the Standard Reference Data Act, the National Bureau of Standards (NBS) of the U.S. Department of Commerce has the primary responsibility in the federal government for providing reliable scientific and technical reference data. The key to the program is decentralization.

The Office of Standard Reference Data of NBS coordinates a complex of data evaluation centers, located in university and other laboratories, as well as within NBS. The data centers compile and critically evaluate numerical data on physical and chemical properties and may also provide services such as answering inquiries for specific data and production of custom bibliographies.

The NSRDS comprises the set of data centers and other data evaluation projects administered or coordinated by the National Bureau of Standards.

The primary aim of this program is to provide critically evaluated numerical data, in convenient and accessible form, to the scientific and technical community of the United States. A second aim is to advance the level of experimental measurements by providing feedback on sources of error in various measurement techniques. Through both these means, the program strives to increase the effectiveness and productivity of research, development, and engineering design.

The technical scope of the program is restricted to well-defined physical and chemical properties of substances and systems that are well characterized. Although this definition leaves some borderline cases, the intent is to concentrate the effort on intrinsic properties that are clearly defined in terms of accepted physical theory. Properties that depend on arbitrarily defined characteristics of the measurement technique are generally excluded. Likewise, materials of uncertain or variable composition are not included. Biological properties and data relating to large natural systems (e.g., the atmosphere, the oceans) also fall outside the program.

In terms of major applications of the outputs, current projects fall into the following categories:

- *Energy and Environmental Data*—This program includes projects dealing with data that have an important application in some aspect of energy R&D or environmental quality improvement. Projects in chemical kinetics, nuclear properties, spectroscopic data, and interaction of radiation with matter are currently incorporated in this program. The output of these projects is particularly important in R&D on new energy sources, environmental monitoring techniques, and prediction of the effects of pollutants introduced into air, water, or land.
- *Industrial Process Data*—Projects dealing with thermodynamic, transport, colloid and surface, and physical properties of industrially important substances are included in this program. Such data have particular application to design of new processes in the chemical and metallurgical industries, optimization of currently used processes, and general productivity enhancement.
- *Materials Utilization Data*—This program covers properties required for material selection and R&D on new materials. The structural, optical, electric, magnetic and mechanical properties of solid materials are included.
- *Physical Science Data*—Projects that involve basic data of broad applicability, or which are associated with an important frontier field of science, are included in this program. Examples are fundamental physical constants, data on fundamental particles, and data relevant to radioastronomy.

The principal output of the program consists of compilations of evaluated data and critical reviews of the status of data in particular technical areas. Evaluation of data implies a careful examination, by an experienced specialist, of all published measurements of the quantity in question, leading to the selection of a recommended value and a statement concerning its accuracy or reliability. The techniques of evaluation depend on the data in question but generally include an examination of the method of measurement and the characterization of the materials, a comparison with relevant data on other properties and materials, and a check for consistency with theoretical relationships. Adequate documentation is provided for the selections of recommended values and accuracy estimates.

As noted earlier, the NSRDS is managed by the Office of Standard Reference Data (OSRD) of the NBS. This office has the responsibility of allocating that part of the NBS budget that is spent on critical data evaluation, both within the NBS technical divisions and through contracts with outside organizations. The staff of the office act as monitors for all supported projects. The management of the publications program of NSRDS is also in the hands of OSRD, and an information service is operated on a limited scale. In addition, OSRD maintains close contact with other data compilation activities, both in the United States and abroad. It attempts, both domestically and internationally, to avoid needless duplication and to encourage coverage of important technical areas. Evaluated data produced under the NSRDS program are disseminated through the following mechanisms:

- *Journal of Physical and Chemical Reference Data*—A quarterly journal containing data compilations and critical data reviews, published for the NBS by the American Institute of Physics and the American Chemical Society. The objective of the journal is to provide critically evaluated physical and chemical property data, fully documented as to original sources and the criteria used for evaluation. Critical reviews of measurement techniques, whose aim is to assess accuracy of available data in a given technical area, are also included. The journal is not intended as a publication outlet for original experimental measurements such as are normally reported in primary research literature, nor for review articles of a descriptive or primarily theoretical nature. Its scope of coverage includes critically evaluated data in the following areas: atomic and molecular properties, chemical kinetics parameters, colloid and surface properties, mechanical properties of materials, nuclear properties, solid state properties, and thermodynamic and transport properties. A handy guide to the contents of this important journal is its *Reprint No. 89* which contains property and author indexes to Volumes 1–5 (1972–1976). The *Data Compilation Abstracts* section in each issue lists and briefly describes selected data compilations and reviews and is one way to keep up with what is being published elsewhere.

- NSRDS-NBS Series—A publication series distributed by the Superintendent of Documents, U.S. Government Printing Office.

- Appropriate publications of technical societies and commercial publishers.

- Response by OSRD and individual data centers to inquiries for specific data.

The most recent publications list of NSRDS can be scanned to find out what is available in printed form. Also, NSRDS publishes an informal newsletter, *Reference Data Report,* which has news and ideas about data centers, publications, meetings, and other activities related to data evaluation and dissemination.

Further information on NSRDS publications, sources of data, and support of data compilation activities can be obtained from: Dr. David R. Lide, Jr., Chief, Office of Standard Reference Data, National Bureau of Standards, Washington, DC 20234, Telephone: (301) 921-2467.

1. Data Centers and Projects

Listed below are titles, locations, and project leaders for all continuing data centers and short-term projects that receive at least a part of their support from the OSRD. A supplementary list of continuing data centers in the United States that are recognized as part of NSRDS (even though they do not receive direct financial support from NBS) follows the project listings in each program area. The source of this tabulation is *NBS Technical Note 947*, January 1977, which provides a status report on the NSRDS and a brief description of each center or project.

Energy Data

Atomic Energy Levels Data Center, Optical Physics Division, NBS, Washington, DC 20234—W. C. Martin.

Transition Probabilities Data Center, Optical Physics Division, NBS, Washington, DC 20234—W. L. Wiese.

Atomic Collision Cross Section Information Center, Laboratory Astrophysics Division, NBS Boulder, Boulder, CO 80302—E. C. Beaty.

Photonuclear Data Center, Center for Radiation Research, NBS, Washington, DC 20324—E. G. Fuller.

X-ray and Ionizing Radiation Data Center, Center for Radiation Research, NBS, Washington, DC 20234—J. H. Hubbell.

Molten Salts Data Center, Rensselaer Polytechnic Institute, Troy, NY 12181—G. J. Janz.

Kinetics of High Temperature Reactions, 3130 Coronado, Santa Clara, CA 95051—Robert Shaw.

Reaction Rate Data for Electronically Excited Atoms and Molecules, ChemData Research, 260 Loma Media, Santa Barbara, CA 93103—K. Schofield.

Table of Isotopes, Lawrence Radiation Laboratory, University of California, Berkeley, CA 94720—C. Michael Lederer.

Fundamental Vibration Frequencies of Molecules, Department of Physical Chemistry, University of Tokyo, Tokyo, Japan—T. Shimanouchi.

Index to High Resolution Spectral Data, Optical Physics Division, NBS, Washington, DC 20324—Paul H. Krupenie.

Molten Salts Data for Energy Storage Application, Rensselaer Polytechnic Institute, Troy, NY 12181—G. J. Janz.

Composite Materials Data for Energy Storage Application, Plastics

Technical Evaluation Center, Picatinny Arsenal, Dover, NJ 07801—H. E. Pebly.

Solid Electrolyte Material Data for Energy Storage Application, University of Utah, Salt Lake City, UT 84112—Gerald Miller.

Metal Properties Data for Energy Storage Application, Cryogenics Division, NBS Boulder, Boulder, CO 80302—H. M. Ledbetter.

The following data centers, not directly under NSRDS program management, also supply evaluated data relevant to this area of interest:

Nuclear Data Project, Oak Ridge National Laboratory, P. O. Box X, Oak Ridge, TN 37830—B. Ewbank.

National Nuclear Data Center, Brookhaven National Laboratory, Upton, NY 11973—S. Pearlstein.

Physical Data Group, Lawrence Radiation Laboratory, University of California, Livermore, CA 94550—Robert F. Howerton.

Environmental Data

Chemical Kinetics Information Center, Physical Chemistry Division, NBS, Washington, DC 20234—R. F. Hampson, Jr.

Radiation Chemistry Data Center, Radiation Laboratory, University of Notre Dame, Notre Dame, IN 46556—Alberta Ross.

Ion Energetics Data Center, Physical Chemistry Division, NBS, Washington, DC 20234—H. M. Rosenstock.

Solubility Data of Inorganic Substances in Water, Part I, Emory University, Atlanta, GA 30322—H. L. Clever.

Solubility Data of Inorganic Substances in Water, Part II, Virginia Polytechnic Institute, Blacksburg, VA 24061—A. F. Clifford.

Infrared Spectral Data for Environmental Applications, The Coblentz Society (care of Chemir Laboratories, 761 West Kirkham, Glendale, MO 63122)—C. D. Craver.

Rate Constant Data on Hydrolysis of Organic Compounds, Stanford Research Institute, Menlo Park, CA 94025—T. Mill and W. Mabey.

Ion-Molecule Reaction Rate Data, Physical Chemistry Division, NBS, Washington, DC 20234—L. W. Sieck.

NMR Data Compilation, Department of Chemistry, Texas A & M University, College Station, TX 77843—B. L. Shapiro.

The following data center, not directly under NSRDS program management, also suplies evaluated data relevant to this area of interest:

API44-TRC Selected Spectral Data, Thermodynamics Research Center, Texas A & M University, College Station, TX 77843—B. J. Zwolinski (see the next text section for details on projects at Texas A & M).

Industrial Process Data

Chemical Thermodynamics Data Center, Physical Chemistry Division, NBS, Washington, DC 20234—D. Wagman.

Thermodynamics Data for Industrial and Municipal Incinerator Processes, Physical Chemistry Division, NBS, Washington, DC 20234—E. Domalski.

Thermodynamics Data on Organic Compounds, Texas A & M University, College Station, TX 77843—B. J. Zwolinski.

Properties of Coal-Derived Compounds, Texas A & M University, College Station, TX 77843—B. J. Zwolinski.

Thermodynamic Properties of Polar Fluids, Heat Division, NBS, Washington, DC 20234—M. Klein.

Cryogenic Data Center, Cryogenics Division, NBS, Boulder, CO 80302—N. A. Olien.

LNG Materials and Fluids Data Book, Cryogenics Division, NBS, Boulder, CO 80302—D. B. Mann.

PVT and Related Thermodynamic Properties of Ethylene, Office of Standard Reference Data, NBS, Washington, DC 20234—H. J. White, Jr.

Thermodynamic Properties of Fluids in the Critical Region, Heat Division, NBS, Washington, DC 20234—M. Klein.

Cryogenic Fluid Mixture Properties, Cryogenics Division, NBS, Boulder, CO 80302—M. H. Hiza.

Excess Property Data for Binary Liquids, Thermodynamic Research Laboratory, Washington University, St. Louis, MO 63130—D. Buford Smith.

Aqueous Electrolyte Data Center, Physical Chemistry Division, NBS, Washington, DC 20234—B. R. Staples.

High Pressure Data Center, Brigham Young University, Provo, UT 84601—H. Tracy Hall.

Alloy Data Center, Metallurgy Division, NBS, Washington, DC 20234—Gesina C. Carter.

Phase Diagrams and Thermodynamic Data for Ternary Copper, University of Wisconsin—Milwaukee, Milwaukee, WI 53201; and L. M. Schetky, International Copper Research Association, New York, NY 10022.

Fluid Transport Properties, Cryogenics Division, NBS, Boulder, CO 80302—H. J. M. Hanley.

Correlation of Thermophysical Property Data of Fluids, University of Maryland, College Park, MD 20742—J. V. Sengers.

Thermal Conductivity, Thermophysical Properties Research Center Purdue University, Lafayette, IN 47906—Y. S. Touloukian (see the following text section for details on projects at this center).

The following data centers, not directly under NSRDS program management, also supply evaluated data relevant to this area of interest:

National Center for the Thermodynamic Data of Minerals, U. S. Geological Survey, Reston, VA 22092—J. L. Haas, Jr.

JANAF Thermochemical Tables, Dow Thermal Research Laboratory, Midland, Michigan 48640—M. W. Chase (see the following text section p. 206 for detail on this project).

International Copper Research Association, New York, NY 10022—M. L. Schetky.

Thermochemistry of Metallurgy, Department of Metallurgy, Massachusetts Institute of Technology, Cambridge, MA 02139—J. F. Elliott.

Contributions to the Data on Theoretical Metallurgy, Thermodynamics Laboratory, Albany Metallurgy Research Center, Albany, OR 97321—S. Hill, Project Director, and A. D. Mah.

Materials Utilization Data

Superconductive Materials Data Center, General Electric Company, Schenectady, NY 12301—B. W. Roberts.

Crystal Data Center, Inorganic Materials Division, NBS, Washington, DC 20234—Helen M. Ondik.

Cambridge Crystallographic Data Centre, Cambridge University, Cambridge, England—Olga Kennard.

Moessbauer Effect Data Center, University of North Carolina, Asheville, NC 28804—John G. Stevens.

Diffusion in Metals Data Center, Metallurgy Division, NBS, Washington, DC 20234—J. R. Manning.

Elastic Constant Data for Metals and Alloys, Cryogenics Division, NBS, Boulder, CO 80302—R. P. Reed.

Optical Properties of Materials, Center for Information and Numerical Data Analysis and Synthesis (CINDAS), Purdue University, West Lafayette, IN 47907—H. H. Li.

Fracture Mechanics Data Handbook, Metallurgy Division, NBS, Washington, DC 20234; and Department of Mechanical Engineering, University of Maryland, R. deWit, NBS, Washington, DC 20234.

Vibrational Force Field Constants for Polyethylene, Polymers Division, NBS, Washington, DC 20234—D. H. Reneker.

Physical Sciences Data

Fundamental Physical Constants, Electricity Division, NBS, Washington, DC 20234—Barry N. Taylor.

Microwave Spectral Tables, Optical Physics Division, NBS, Washington, DC 20234—Frank J. Lovas.

Berkeley Particle Data Center, Lawrence Berkeley Laboratory, University of California, Berkeley, CA 94720—Arthur R. Rosenfeld.

Some details on the centers at Texas A & M and Purdue University are presented in the next sections.

2. Thermodynamics Research Center

One of the best known of the data centers affiliated with NSRDS is the Thermodynamics Research Center (TRC), Texas A & M University, College Station, TX, which functions under the direction of Dr. Bruno J. Zwolinski. This center is notably strong in data for organics; inorganics are also covered.

Two major efforts underway at TRC are described below:

a API Research Project 44 (APIRP 44), "Selected Values of Properties of Hydrocarbons and Related Compounds," is one of the most complete compilations of physical, thermodynamic, and spectral data available on these compounds. The project began in 1942 at the NBS with financial support from the American Petroleum Institute (API), and under the direction of Professor F. D. Rossini. In 1950 it was moved to the Carnegie Institute of Technology, Pittsburgh. In 1961, Professor Bruno J. Zwolinski assumed the directorship, and the project was moved to Texas A & M University. Dr. Zwolinski is aided by a staff of professional and support personnel. Since 1966 API Research Project 44 has been on a self-supporting basis, with an API Advisory Committee of data specialists from the petroleum and petrochemical industries providing scientific and technical guidance. The APIRP 44 continues to serve as the central agency of the API for collection, evaluation, publication, and distribution of numerical data on the physical, thermodynamic, and spectral properties of all classes of hydrocarbons, together with certain classes of sulfur and nitrogen derivatives of hydrocarbons present in petroleum.

b The "Thermodynamics Research Center Project" (TRCDP) began in 1955 at the Carnegie Institute of Technology, under auspices of the Manufacturing Chemists Association. Sponsorship changed July 1, 1966 to the Texas A & M Thermodynamics Research Center. Since 1963 this project has received continuing support from the OSRD of the NBS. The principal objective is the preparation, publication, and distribution of spectral data and data on physical and thermodynamic properties of all classes of organic compounds (except hydrocarbons and related compounds covered by APIRP 44), as well as industrially important nonmetallic inorganic compounds.

This data bank, too, is one of the most complete compilations of spectral and physical and thermodynamic data to be found on these compounds.

The *Comprehensive Index of API44—TRC Selected Data on Thermodynamics and Spectroscopy* (13-1) is a useful aid in locating data for about 10,000 compounds in the 12 publications of TRC. The categories of data found are IR, UV, Raman, Mass, NMR, and physical and thermodynamic properties from the beginnings of the project, more than 30 years ago, through 1973.

The APIRP 44 and the TRCDP tables include data for many chemical compounds of interest and importance to the chemical and petroleum industries, and to science in general. The tables are based on the best experimental data available. This information is critically evaluated and presented in a convenient, looseleaf form for ready use. Furthermore, data are estimated in temperature and pressure ranges that are not easily accessible in the laboratory. Data are often estimated for new compounds not yet synthesized but that may possess desirable properties.

3. Center for Information and Numerical Data Analysis and Synthesis

Another major data analysis center associated with NSRDS is the Center for Information and Numerical Data Analysis and Synthesis (CINDAS) located at Purdue University, West Lafayette, IN. The center was founded in 1957 by Dr. Y. S. Touloukian, who is currently the director.

One of the centers located within CINDAS is the Thermophysical Properties Research Center (TPRC). Information and data on more than 60,000 different substances and materials are currently coded for 14 properties:

Thermal conductivity
Accommodation coefficient
Thermal contact resistance
Thermal diffusivity
Specific heat at constant pressure

Viscosity
Emittance
Reflectance
Absorptance
Transmittance
Solar absorptance to emittance ratio
Prandtl number
Thermal linear expansion coefficient
Thermal volumetric expansion coefficient

The Properties Research Laboratory, an affiliate of CINDAS, measures the thermal, electrical, and optical properties of materials over wide temperature ranges.

COMMITTEE ON DATA FOR SCIENCE AND TECHNOLOGY

At the international level, the Committee on Data for Science and Technology (CODATA) was created by a special committee of the International Council of Scientific Unions to aid in international coordination of data compilation, retrieval and evaluation. Material available from CODATA includes its *Bulletin* which contains proceedings of CODATA conferences, reviews of new data publications, and articles on methodology of data evaluation. Another product of chemical interest from CODATA is their *Recommended Key Values for Thermodynamics* (see p. 204).

CODATA also compiled a *Compendium* (13-2), published in 1969, which lists, among other things, continuing numerical data projects and their publications. An attempt has been made to update and expand the *Compendium* through the *Bulletin*, but it is now being completely revised to reflect the many changes since 1969.

Newly established by CODATA is a "World Data Referral

Centre" to guide users on availability and sources of needed data. The most recent address for CODATA is: CODATA Secretariat, 51 Boulevard de Montmorency, 75106 Paris, France.

JOURNAL OF CHEMICAL AND ENGINEERING DATA AND OTHER JOURNALS

The *Journal of Chemical and Engineering Data (JCED)* published since 1956 by the American Chemical Society, although not related to NSRDS, is a good example of a current and original source of reliable physical property data for both pure compounds and mixtures.

A quick scan of indexes to this publication and of the tables of contents of its more recent issues can sometimes produce more useful information than an equivalent amount of time spent perusing the printed subject indexes to *CA*. No search for the most recent physical property data for specific compounds can be considered complete without consulting *JCED* either directly or through the use of indexing and abstracting services.

In addition to containing experimental or derived data relating to pure compounds or mixtures covering a range of states, *JCED* includes:

1 Papers based on published experimental information that have made tangible contributions.

2 Experimental data that aid in identifying or utilizing new compounds.

3 Papers relating to newly developed or novel syntheses of organic compounds and their properties.

JCED sets a high standard. Its guide for authors specifies that experimental methods should be referenced or described in enough detail to permit duplication of the data by others familiar with the field. Published or standardized procedures and their simple modifications need not be described, but a readily available reference should be cited. The data should be presented with such precision that information may be easily obtained from within the paper, within the stated limits of uncertainty of the experimental background.

This journal is cited as just one example of many in the field.

Another leading example is the *Journal of Chemical Thermodynamics*. Others are published by the American Institute of Chemical Engineers, the American Institute of Physics, the American Chemical Society, and other sources.

Note that both scientific journals and trade magazines can be rich sources of data, but often different kinds of data.

Articles in scientific journals concerned primarily with preparative chemistry are good sources of property data, primarily "basic" properties such as melting point or refractive index. If access to such articles is made through abstracting and indexing services, the chemist must note that these properties are not necessarily indexed or abstracted, since their presence is often implicitly assumed.

Trade-oriented magazines frequently have articles with information on properties affecting end use or applications, such as data relating to handling or processing.

TRADE LITERATURE

The trade literature of chemical manufacturers often gives full and accurate treatment of physical and chemical properties, especially those values that relate to use and to safe handling. Original data, which may not be available elsewhere, are frequently presented here. In using this literature, the chemist needs to make careful note of the grade or purity for which the properties are specified.

For the most recent data, and that not found in trade

literature, the chemist should write or telephone the manufacturer's technical representative who specializes in the product or product line.

BULLETIN OF CHEMICAL THERMODYNAMICS

A key to recent thermodynamic data and literature is the annual *Bulletin of Chemical Thermodynamics*, edited by Professor Robert D. Freeman, Oklahoma State University, Stillwater, OK 74074. The *Bulletin* has three main sections: index, bibliography, and reports.

The bibliography is a listing of papers with chemical thermodynamic content published during the preceeding year. The reports section provides terse summaries of studies completed but not yet published. Some 35 countries are covered. Actual data are not given (except for such information as range of variables), but there is sufficient detail to permit users to decide whether to make contact with the cited investigators. The index provides access to the other two sections by substance or property. Other information included is a calendar of meetings, a list of pertinent books, and beginning in 1978, review articles.

The *Bulletin* is an international effort, prepared under auspices of the International Union of Pure and Applied Chemistry (IUPAC).

HANDBOOKS

The most heavily used sources of physical property data are now, and have been for many years, the *CRC Handbook of Chemistry and Physics* (13-3) and the *Chemical Engineers Handbook* (13-4). These provide rapid, convenient, desk-top access to many different kinds of data and can be purchased for prices within the budgets of most practicing chemists and engineers.

These and other desk handbooks usually contain more than just physical properties—a feature which further enhances the value of handbooks. For example, the *Handbook of Chemistry and Physics* includes such other material as an extensive section on mathematical tables, information on sources of critical data, and rules for nomenclature of organic chemistry. The *Chemical Engineers Handbook* includes a section on mathematical tables, data on materials of construction and corrosion, and information on cost and profitability. Another handy desk reference volume is the *Chemist's Companion* (13-5).

Chemists will find the ninth (1976) edition of *The Merck Index* (13-6) exceptionally useful. This has descriptive information on 10,000 chemicals, drugs, and biologicals arranged alphabetically by generic or nonproprietary name. Although the publisher is a major pharmaceutical company, coverage is across the board for the most important chemicals of many types. The volume includes 500 organic name reactions, a comprehensive cross-index of 50,000 synonyms, a formula index, and a variety of tabular information. Chemical Abstracts Service Registry Numbers are given when available, and there are carefully selected references to key journal and patent literature for many of the chemicals listed. More than 50% of the chemicals are illustrated with stereochemical structural formulas. For many chemicals included there is information on general, medical, or veterinary uses as well as toxicity.

The volumes cited above are examples of the many kinds of handbooks of value to chemists. The CRC Press, Inc., Cleveland, OH, probably the best known publisher of handbooks of interest to chemists and engineers, recently (1977) announced a composite index that provides access to data from the entire collection of almost 50 CRC handbooks. (13-7)

Chemists and engineers should use the data in desk handbooks—although valuable, convenient, and indispensable from a practical, day-to-day working standpoint—only with some appropriate cautions:

1 New data are constantly being generated and published in journals and other sources. Handbooks, especially those published in conventional bound form, cannot (and do not) claim to provide the most recent data. Thus desk handbooks can be described as incomplete and out-of-date but not necessarily incorrect.

2 Tables printed in many handbooks have not been subjected to direct and recent critical review and evaluations. Hence the data are not necessarily the "best" data—an important factor in many chemical investigations.

3 Handbook data are usually based on, or "copied" from, other sources. Typographical and other errors are more likely to occur than in original sources.

4 Details of conditions under which data were originally generated are not always given in handbooks. Limitations are not always evident. If complete references to the original sources of the data are also missing, this further complicates attempts at data evaluation by handbook users.

5 The indexes may not be sufficiently detailed to locate the desired information without considerable "digging." Occasional outright loss of data can occur because of this difficulty. Handbook editors, especially in newer editions, are working to improve access.

6 Some chemical engineers believe that single values are not too useful for engineering calculations and that curves of values, not just single points, are desirable. Tabular values are, however, frequently not data points, but rather "smoothed" and interpolated values.

7 The latest editions of some handbooks omit useful data found in earlier editions. It is worthwhile to keep on hand or consult at least one prior edition.

Even with these and other limitations, the chemist or engineer will probably continue to make heavy use of handbooks as a convenient source of information for investigations in which the highest degree of up-to-dateness, accuracy, or precision is not essential. Handbooks, such as those mentioned, serve as a good starting point when the latest, best, and most complete data are not needed.

INTERNATIONAL CRITICAL TABLES

One of the most extensive compilations of physical property data is the *International Critical Tables* (13-8), often referred to as the *ICT*. This eight-volume set, published in the late 1920s and early 1930s, is now out of print, although still on the shelves of many chemistry libraries.

The chemist will find here some still valid data, but the compendium has been largely superseded by newer works. Today, *ICT* can be regarded as a "court of last resort," with much of its utility limited to unusual or "exotic" property data not likely to be found elsewhere.

EXAMPLES OF OTHER REFERENCE SOURCES

Some key reference sources are listed below. Most descriptions are as given in *NBS Special Publication 454* (13-9).

S. W. Benson, *Thermochemical Kinetics—Methods for the Estimation of Thermochemical Data and Rate Parameters*, 2nd ed., New York, Wiley, 1976. Molecular approach to techniques for rapid and relatively quantitative estimation of thermochemical data and reaction-rate parameters for chemical reactions in the gas phase. An appendix contains a collection of tabulated group values for different classes of organic compounds and thermochemical data for selected elements, radicals, and molecules. Also included are tables for estimates of $S°$ and $C_p°$ contributions from vibration and torsional degrees of freedom as a function of temperature, frequency, and barrier.

F. R. Bichowsky, and F. D. Rossini, *The Thermochemistry of the Chemical Substances*, Reinhold, New York, 1936. Although outdated, still provides useful references to older thermochemical literature. Tabulated are $\Delta H_f°$ values for the elements and their compounds, with data for carbon-containing compounds being terminated at two carbon atoms. The data pertain to a temperature of 18°C and to diamond, rather than graphite, as the standard state for carbon.

CODATA Recommended Key Values for Thermodynamics 1975 (*CODATA Bulletin* No. 17 (1976) and Tentative Set of Key Values for Thermodynamics: Part V, CODATA Special Report 3, September 1975). Recommended values for the quantities $\Delta H_f°$ (298.15 K), $S°$ (298.15 K), and $H°$ (298.15 K) − $H°$ (O K) for 102 of the thermochemically more important elements and compounds, including some aqueous species. These bulletins supersede earlier CODATA reports of this group. The recommended values are not fully consistent with any previously published thermodynamic tables, but the values are intended to form the basis of future generations of compilations (see also annotation on compilation by Parker, Wagman, and Garvin).

J. D. Cox, and G. Pilcher, *Thermochemistry of Organic and Organometallic Compounds*, Academic, London, New York, 1970. A critical compilation of thermochemical data for the title field published since 1930. The enthalpies of formation of some 3000 substances are listed, with estimates of error. Where enthalpies of vaporization are known or can be reliably estimated, these are listed and in these cases the enthalpies of formation of both gaseous and condensed phases are given. Extensive introductory material presents experimental procedures for reduction of experimental data of the type found in the book. Applications of thermochemical data are given, and there is a section on methods of estimating enthalpies of formation of organic compounds.

C. H. Horsley, Ed., *Azeotropic Data—III*, American Chemical Society, Washington, DC, 1973. Data on azeotropes, nonazeotropes, and vapor-liquid equilibria for more than 17,000 systems. No attempt has been made to evaluate accuracy of the data, most of which are from the original literature. This volume is a revision of *Azeotropic Data I and II* and includes new data collected since 1962.

G. J. Janz, *Thermodynamic Properties of Organic Compounds Estimation Methods, Principles, and Practice* (rev. ed.), Academic, New York, 1967. Computation of thermodynamic properties such as heat capacities, entropies, enthalpies, and Gibbs energies by statistical mechanical methods and by methods of structural similarity, group contributions, group equations, and generalized vibrational assignments. The chemical properties enthalpy of formation and enthalpy of combustion are treated in terms of bond energies and group increments. Some 78 tables are given of increments, group contributions, and bond contributions as specifically needed for estimation of particular properties.

T. E. Jordan, *Vapor Pressure of Organic Compounds*, Interscience, New York, 1954. A comprehensive compilation of vapor pressure data for organic compounds. Included are tables on the hydrocarbons, alcohols, aldehydes, esters, ketones, acids, phenols, and metal organic compounds. Data for each compound are shown in graphical form, i.e., vapor pressure as a function of temperature. References to the data sources in the literature are given.

J. A. Larkin, Ed., *International Data Series, Series B, Data on Aqueous Organic Systems*, National Physical Laboratory, Teddington, Middlesex, U.K. A relatively new series for dissemination of selected data on mixtures. Similar in format, scope, and policies to Series A *(Thermodynamic Properties of Binary Systems of Organic Substances)*.

W. F. Linke, and A. Seidell, *Solubilities: Inorganic and Metal-organic Compounds—A Compilation of Solubility Data From the Periodical Literature*, Volume 1: A-Ir, Volume II: K-Z Volume I: D. Van Nostrand, Princeton, NJ, 1958; Volume II: American Chemical Society, Washington, DC 1965. Comprehensive compilation of mostly unevaluated solubility data for inorganic and metal-organic compounds. Both aqueous and nonaqueous solvent systems are included. References are given to the data sources.

V. B. Parker, D. D. Wagman, and D. Garvin, *Selected Thermochemical Data Compatible With the CODATA Recommendations*, National Bureau of Standards Report No. 75-968, Washington, DC, 1976. Selected thermochemical properties, ΔG_f°, ΔH_f°, S°, C_p° (all at 298.15 K), ΔH_f° (O K), and H (298 K)—H (O K), are given for 384 substances (almost entirely inorganic) including many of the more commonly encountered aqueous species. The selected values are intended to be compatible with the current CODATA recommendations on key values for thermodynamics. (see *CODATA Recommended Key Values for Thermodynamics*).

H. Stephen and T. Stephen, Solubilities of Inorganic and Organic Compounds, (5 vols.) Pergamon, London, 1963. Selection of data on the solubilities of elements, inorganic compounds, and organic compounds in binary, ternary, and multicomponent systems. References are given to sources of data in the literature. The data are unevaluated.

T. S. Storvick, and S. I. Sandler, Eds., *Phase Equilibria and Fluid Properties in the Chemical Industry—Estimation and Correlation*, American Chemical Society, Washington, DC, 1977. A symposium volume containing state-of-the-art reviews.

D. R. Stull, "Vapor Pressure of Pure Substances. Organic Compounds," *Industrial and Engineering Chemistry*, **39**, 517–540 (1947); "Vapor Pressure of Pure Substances. Inorganic Compounds," *Industrial and Engineering Chemistry*, **39**, 540–550 (1947). Evaluated vapor pressure data on over 1200 organic and 300 inorganic compounds.

D. R. Stull and H. Prophet, *JANAF Thermochemical Tables*, 2nd ed. National Standard Reference Data Series, National Bureau of Standards, Report No. 37, U. S. Government Printing Office, Washington, DC 1971. The JANAF (Joint Army-Navy-Air Force) tables contain data on thermodynamic properties for over 1000 chemical species. This is a particularly good source for workers in inorganic chemistry; some simple organics are also included. Some of the data supplement, parallel and reinforce those found in NBS-TN-270. Supplements are published in the *Journal of Physical and Chemical Reference Data* beginning in 1974.

D. R. Stull, E. F. Westrum, and G. C. Sinke, *The Chemical Thermodynamics of Organic Compounds*, Wiley, New York, 1969. Monograph divided into three parts. The first gives the theoretical basis and principles of thermodynamics and thermochemistry, some experimental and computational methods used, and some applications to industrial problems. The second part gives thermal and thermochemical properties in the ideal gas state from 298 to 1000 K. In this section the sources of data are listed and discussed, and standardized tables are presented for 918 organic compounds. Values of $C_p°$, $S°$, $-(G - H°_{298})/T$, $H°$, $H°_{298}$, $\Delta H_f°$, $\Delta G_f°$, and log K_p are given at 100 K intervals. In the third section are listed selected values of enthalpy of formation, entropy, and consistent values of $\Delta G_f°$ and log K_p of organic compounds at 298 K. More than 4000 compounds are listed, including a few inorganic compounds. A chapter briefly discusses methods of estimating thermodynamic quantities.

D. D. Wagman, W. H. Evans, V. B. Parker and (in various individual parts) I. Halow, S. M. Bailey, R. H. Schumm, K. L. Churney, *Selected Values of Chemical Thermodynamic Properties*, National Bureau of Standards Technical Note 270, U.S. Government Printing Office, Washington, DC 1965. A revision of the well-known NBS Circular 500 Part I, issued in parts as segments of the work relating to selected sequences of elements are completed. The following parts have been issued as of 1977:

270–1 Tables for the first 23 elements in the standard order of arrangement.

270–2 Tables for elements 24–32 in the standard order of arrangement.

270–3 Tables for the first 34 elements in the standard order of arrangement. This table includes thermochemical data for compounds containing one or two carbon atoms. This supersedes Technical Notes 270–1 and 270–2.

270–4 Tables for elements 35–53 in the standard order of arrangement.

270–5 Tables for elements 54–61 in the standard order of arrangement.

270–6 Tables for the alkaline earth elements, elements 92–97 in the standard order of arrangement.

270–7 Tables for the lanthanide (rare earth) elements, elements 62–76 in the standard order of arrangment.

The remaining elements are to be covered in additional parts. This is the most comprehensive recent compilation in English of critically evaluated thermochemical data at 298.15 K for inorganic substances. All inorganic substances and organic substances containing two carbon atoms or fewer per molecule are included if thermodynamic data exist for calculating any of the properties tabulated. The coverage, when complete, will be approximately 12,000 substances.

EVALUATING DATA FROM CONFLICTING OR UNEVALUATIVE SOURCES

What if rigorously evaluated data such as that provided by NSRDS or other sources is not available? What if adequate verification of the data in the chemist's laboratory is not possible?

In such cases the chemist may need to make his own relatively subjective assessment of published data on physical and chemical properties. Data from several sources may coincide or may vary significantly. Data from a single source may be suspect. In determining reliability of data found in the literature, the chemist needs to consider factors such as these examples:

Is the chemical identity of the substance unambiguously specified? This is a basic starting point, because it helps avoid pitfalls due to variations in nomenclature and in the many commercial trade names and grades.

Is there agreement between several independent sources? Agreement between two or more modern, accurate determinations is usually considered to be a good criterion of accuracy, especially if deviations from the average are about 3% or less. When there is lack of agreement, one choice is to take the safest, most conservative value. For example, in considering different flash-point values, the chemist could select the lowest value to provide the greatest possible safety margin (*note the importance of checking in several different sources for values of the same constant*).

Is the source of the data a paper written to determine specific physical constants rather than a paper in which the constant is determined only incidentally as in the course of a laboratory preparation? The former is a more reliable source.

How recent is the source? Recently determined values (and the more recent the better) are usually preferred over older values because of improvements in techniques and apparatus and advances in knowledge—if everything else is equal. But this does not mean that older sources of information should be totally discounted or neglected.

Is the source a specialized book such as a monograph on specific compounds or classes of compounds? Such books are frequently preferred as sources of physical property data over more general treatises, which are less specialized.

Do the sources being compared refer to the same original source for their data? In comparing secondary (i.e., unoriginal or secondhand) sources, the chemist should beware of the possibility that the secondary sources are all based on the same original source, which may be erroneous.

Is the author (investigator) a specialist in the chemical or property being studied, and what is that person's reputation for good work?

What is the reputation of the laboratory or research center where the work was done? For example, high confidence is placed in work done at the U.S. National Bureau of Standards.

What is the reputation of the publisher or publication? For example, values appearing in a source such as the *Journal of the American Chemical Society* would ordinarily be given more credence than a value appearing in a journal of unknown or questionable reputation.

Do the data appear in tables or graphs in which even a single error is detected? If the answer is yes, such data need to be looked on with extraordinary scrutiny.

How much experimental detail is given? Inclusion of full detail enhances confidence in the results. Lack of detail diminishes confidence. Some specifics for which detail is helpful and which should be looked for include:

a Purity of material on which the determination is made.

b Source of the material (supplier).

c Type of apparatus used in making the determination and precision of that apparatus.

d Reliability and reputation of methods used.

e Limits of error or confidence limits.

The lack of such information could reduce or even entirely negate the placing of confidence in data reliability. Inclusion of such information increases confidence and facilitates evaluation.

Are all pertinent facts reported, including those unfavorable to the author's position or theory? To what extent are opposing interpretations and views included?

Do the results appear to be overprojected and overextended?

Are the data internally consistent?

Is the work written so that it can be correlated, repeated, or verified by others?

Are interpretations clearly labeled as such?

Is there an attempt to evaluate and assess the reliability of the results critically?

DETERMINING OR ESTIMATING PROPERTIES

If a careful search of literature and other related information sources does not reveal needed data of high reliability, one or more of the following steps can be taken:

1 The chemist may find it quicker, easier and more accurate to determine the data in the laboratory—if the required instrumentation and expertise are available—than to spend additional days searching the literature. This decision is a trade-off, based on estimated time and costs involved, as well as importance of the data. Reliable experimental determinations are ordinarily the choice over estimating or predicting, if time and funds permit.

2 The data can be obtained using a computer program, which may or may not have some predictive capabilities. Examples of programs (or sources) include:

a FLOWTRAN, available through Monsanto on a license basis.

b The Chemical Thermodynamic and Energy Release Evaluation Program (CHETAH) developed by a task group of the

American Society for Testing and Materials (ASTM), Philadelphia, PA. It helps predict key thermodynamic properties. As noted on p. 172, CHETAH can be helpful in identifying the relative hazard potential of chemicals.

c CHEMTRAN®, a product of the ChemShare Corp., 2500 Transco Tower, Houston, TX 77056.

d "PPDS" (Physical Property Data Service) developed by the British Institution of Chemical Engineers.

Each of the above has unique capabilities and emphasis; the developers can be contacted directly for details.

There are many other computer programs which could be used. Some are listed in the *Chemical Engineering* compilation "Computer Programs for Chemical Engineers." This was originally published in 1973 (Aug. 20 and Sept. 17) and is still available as the magazine's reprint No. 191. A completely revised update appeared in 1978. The 1973 compilation makes reference to, but does not include for obvious reasons, the seven volumes of computer programs assembled by the CACHE (Computer Aids for Chemical Education) Committee.

Additionally, under the leadership of Dr. Buford D. Smith, the Thermodynamics Research Laboratory at Washington University, St. Louis, MO has as a research objective "enhancing engineering technology by development of improved correlation and prediction methods for thermodynamic properties of liquids and liquid mixtures." The work being done here is highly regarded.

Future sources may include "Project Evergreen" which at this writing is a proposal for a joint industry-government project to support a data base for chemical engineering design calculations. The NBS Office of Standard Reference Data initiated this proposal.

3 It may be possible to infer roughly what the property might be by: (a) studying data for similar classes of related compounds; and/or (b) examining related data for the same compound; and/or (c) studying other compounds in the same homologous series.

4 The chemist can call or write the NSRDS either at their headquarters or at one of their specialized data centers.

5 The chemist can use one of the books written to aid in estimation of properties. The most recent of these is the highly regarded work by Reid, Prausnitz, & Sherwood (13-10). This critical review of the most reliable currently used estimation programs includes specific recommendations. An appendix is a data bank with property values for over 450 pure chemicals. Other books which would help can be selected from the list in an earlier section of this chapter.

14 CHEMICAL MARKETING AND BUSINESS INFORMATION SOURCES

Chemical marketing and business information is important to almost all chemists.

Particularly in industry, the chemical researcher is now being brought out of the relative anonymity and alleged "ivory tower" of the laboratory and into the main stream of the business world. The contemporary researcher is accepted as a full partner by his marketing counterpart. Technological and marketing decision making are highly interdependent. Hence it is not surprising that the boundary between marketing information sources and technical or scientific information sources is "blurred," as previously noted, and that the two types of sources are complementary.

Thus important scientific information may appear for the first time in chemical marketing or business sources. Similarly, scientific and technical literature can provide important leads to chemical marketing specialists. (See also Chapter 4.)

This chapter skims the surface of what is available and how to access it. For more detail, the reader should consult books in the field, such as Giragosian's classic *Chemical Marketing Research* (14-1).

USING MARKETING DATA

As compared to most scientific or technical data, some marketing data, especially estimates or projections into the future, are relatively "soft" or less certain. Laboratory verification of marketing data is not possible. As the chemist becomes more and more involved with the business aspects, he can expect to make increasing use of the tools mentioned in this chapter. In so doing, he should call on the expertise of the chemical marketing specialist, especially when there are questions of interpretation or extrapolation to the future.

INFORMATION ABOUT MANUFACTURERS

In addition to determining who makes a product, the chemist may be interested in chemical manufacturers for other reasons:

a If a company is a possible employer.

b If a company is being considered for personal or other financial investments.

c If the company represents present or potential competition.

d If a joint research or business venture is being contemplated.

e If the company is a potential source for a research grant or other funding.

The information about chemical manufacturers that is usually found most valuable in answering the above and other questions includes:

a Products made by the company.

b Locations (principal offices and plants).

c Names, titles, and backgrounds of officers and other executives.

d Historical background, present status, and future plans.

e Financial data such as sales and profits, preferably broken out by major product line.

A variety of information sources is available to help answer these and other questions. For example:

a Reports issued by companies to stockholders, especially annual reports. These have information about present and historical financial status, new products and other major research achievements, names of key officials, principal locations, and other information.

b Reports filed by companies with the Securities and Exchange Commission (SEC) in Washington, DC. These filings are required by law for all companies that are publicly owned. Most reports filed with the SEC are readily available at SEC offices, directly from the company, or from private organizations which provide this kind of material, for example, Disclosure, Inc., 4827 Rugby Ave., Bethesda, MD 20014. These reports are important because they often contain more detailed financial and marketing information than is found in the annual report issued to stockholders.

c Reports from stockbrokers, and handbooks issued for the financial and investment community (see refs. 14-20, 21, 22). This material is useful in obtaining a description and evaluation of the financial picture of the company—past, present and future—but it also frequently contains other information.

d Patents assigned to the organization and other publications by persons connected with the organization. This helps indicate areas of interest as well as strength of the technical effort.

e News about the company as reported in chemical news magazines and general business and trade periodicals.

f Buyer's guides and other tools discussed in the following sections.

BUYER'S GUIDES AND RELATED TOOLS

Finding out who makes a chemical of interest is one of the most frequent informational needs of the chemist or engineer. Fortunately, a number of useful tools are available.

These include the *SRI* International *Directory of Chemical Producers—U.S.A.* (14-2). This volume is far more than just a buyer's guide. In many cases (for over 200 major products), information is given on capacities at specific plant locations, and in some cases, process or route used is also given. Access is threefold: by product (who makes what, where, how much, and by what route, if known); by company (divisional structure, plant locations, and product produced at each site); and by region (what companies and plants are within specific zip code locations). The location feature is helpful to chemical salespersons and to others concerned with business aspects of the chemical industry (e.g., locating potential sales prospects within a geographical region or locating the nearest contact point or plant of the manufacturer to help expedite delivery).

The annual directory is kept up to date by two cumulative supplements. These note changes in company names and addresses, corporate structure, products produced at each plant location, and mergers and acquisitions. The supplements also contain information on new plants planned or under construction. Subscribers can contact the publisher directly with any questions about listings.

The current edition of the directory is an invaluable source of information to all concerned with the United States chemical industry. Earlier noncurrent editions should be kept because they provide a good historical reference to the state of the chemical industry each year. As indicated in the title, coverage is limited to the United States.

Primary emphasis is on commercial chemicals and on actual producers (distributors are not included). Commercial quantity is defined as "exceeding 0.5 ton (1,000 pounds) or U.S. $1,000 in value annually." For products made in other countries, chemicals made in developmental or research quantities, and names of distributors, other directories need to be consulted. The chemist will find this (and all other buyer's guides) occasionally incomplete because of frequent additions to and deletions from product lines throughout the chemical industry.

Among the other buyer's guides available, the two best known are:

a *Chemical Week Buyer's Guide Issue* (14-3)—This annual publication includes "major producers and sources of supply for more than 6,000 products." There are also addresses and telephone numbers for company offices, and there is a listing of trade names. Similar information is given for companies that provide packages and containers for the shipping of chemicals.

b *OPD Chemical Buyer's Directory* (14-4)—Like the above, this annual volume lists sources of supply for "chemicals and related process materials."

Neither of these publications contains all the information found in the SRI *Directory*, nor is either as complete or up-to-date. But they are much more widely available and are heavily used.

Some chemicals are made in small quantity and for this, or other reasons, may not be listed in the more widely used guides. The best list for such chemicals is *Chem Sources—U.S.A.* (14-5).

A number of specialized guides are available to assist in purchase or identification of chemicals and formulated products intended for specific end uses. There are also special guides for laboratory and other kinds of equipment. Some examples include:

a *Soap Cosmetics Chemical Specialties Blue Book Issue* (14-6)

b *Lockwood's Directory of the Paper and Allied Trades* (14-7)

c *Farm Chemicals Handbook* (14-8) Includes plant foods (fertilizers) and pesticides. Considerable other information given beyond that needed for purchasing.

d *Laboratory Guide to Instruments, Equipment and Chemicals* (14-9)

e *Chemical Engineering Equipment Buyer's Guide* (14-10)

INTERNATIONAL BUYER'S GUIDES

Buyer's guides, such as those previously mentioned, cover primarily United States manufacturers. Because chemistry is international, the chemist should be aware of guides and directories covering other countries. For example:

a *Japan Chemical Directory* (14-11)

b *Chem Sources—Europe* (14-12)

c *Chemistry and Industry Buyer's Guide* (14-13)—A publication of the Society of Chemical Industry, includes information on chemicals and manufacturing plant and laboratory equipment.

d *European Chemical Buyer's Guide* (14-4)—A comprehensive, easy-to-use guide for purchasers of chemicals in Europe, contains information about 4000 suppliers and an alphabetical listing of about 7000 chemicals.

e *The Worldwide Chemical Directory* (14-15)—A worldwide "address book" of chemical industry; includes addresses, telephone numbers, and brief descriptions of activities; contains four sections: list of companies, trade and professional associations, chemical plant contractors, and chemical tanker and storage companies.

Scheduled for 1978 publication is SRI's *Directory of Chemical Producers—Western Europe*. This should be an extremely valuable tool if depth and quality of coverage approximate that of the United States equivalent previously noted.

See also the discussion on Chemical Data Services, beginning on p. 225.

MORE GENERAL GUIDES

On occasion, the chemist or engineer needs to identify sources of equipment or materials not necessarily related directly to chemistry. For these occasions, a comprehensive overall guide for United States manufacturers is *Thomas Register* (14-16). The VSMF files noted on p. 224 may also be useful.

OTHER SOURCES FOR LOCATING CHEMICALS

Chemists who cannot locate sources for chemicals of interest in buyer's guides need not despair. Several options remain open.

Some companies either specialize in stocking hard-to-find chemicals or will consider making these on request-in both cases perhaps in small (research) quantity. Examples include:

1 Aldrich Chemical Co., 940 W. Saint Paul Ave., Milwaukee, WI 53233. Primarily organic and biochemicals: off-the-shelf, custom synthesis service or in larger-than-laboratory quantities.

2 Eastman Organic Chemicals, 343 State St., Rochester, NY 14650.

3 Alfa Products Division of Ventron Corp., Beverly, MA 01915, Specializes in inorganics.

Additionally, there is a directory (14-17) that lists companies which have capability in custom processing.

Finally, the chemist can contact manufacturers of related compounds or of precursors to see if they can provide the chemicals needed.

PRICES

The best regularly published source of prices for chemicals, specifically those used in commerce on a large scale, is the weekly *Chemical Marketing Reporter*. This is an extensive listing, but it is not all inclusive. Even if a chemical is not listed here, it may still be available for purchase. The buyer's guides need to be consulted, as previously recommended.

Because prices depend on quantity, location of buyer and manufacturer, and other variables, an individualized quotation from the manufacturer is desirable for accurate pricing data.

GENERAL CHEMICAL BUSINESS INFORMATION

To help the chemist or engineer locate the latest chemical business information, several good abstracting and indexing services and other tools are available.

Mentioned earlier in this book is *Chemical Industry Notes*, a significant tool in this field published by Chemical Abstracts Service.

Another important service is *Promt* (formerly *Chemical Market Abstracts* and *Equipment Market Abstracts*, now combined) and its associated subsections known as *Predi-Briefs*. This tool is distinguished by in-depth abstracting and versatile indexing access. On-line retrieval is possible through the facilities of Lockheed Information Systems. *Promt*, and other business information tools of value, are published by Predicasts, Inc., 200 University Circle Research Center, 11001 Cedar Ave., Cleveland, OH 44106.

The *Business Periodicals Index* is simply an index (no abstracts), and it is not specifically oriented to chemistry. But it does refer to many articles of chemical business interest, is easy to use, and can be found in many libraries. The publisher is H. W. Wilson Co., 950 University Ave., New York, NY 10452.

Initiated in mid-1976, *Chemical Products Synopsis* is a reporting service on 200 individual major chemical commodities. The succinct reports provide general marketing information, including present situation, short- and long-term outlook, pricing, producers, uses, and brief information on the marketing and environmental aspects. The publisher, Mannsville Chemical Products, is located at Mannsville, NY, 13661.

CHEMICAL ECONOMICS HANDBOOK

For business, marketing, and some technical information on almost all major commercial chemical products and product groups, the *Chemical Economics Handbook (CEH)*, published by SRI International, is a key source of enormous value.

The *CEH* program has been a continuing research activity since 1950, devoted to development of detailed information about the economic progress of the United States chemical industry. The objective of the program is to serve participants' needs to be informed accurately and currently about the present and future status of raw materials, primary and intermediate chemicals, and product groups (such as plastics and fertilizers). Program emphasis is on providing insight into future markets—both in terms of technological requirements of future demand and quantities of chemicals that will be consumed. Background information is also supplied on economic aspects of the chemical industry as a whole, on chemical consuming industries, and on national economic trends.

The *CEH* program is sponsored by subscriptions from

over 300 corporations and other organizations. Research results are available to program participants in the following ways:

- A loose-leaf set of the *CEH*, featuring comprehensive product reports and statistical data sheets.
- Monthly installments of new and revised material to be added to *CEH* volumes.
- Bimonthly issues of the *Manual of Current Indicators* which update the statistical data series in the body of *CEH*.
- An inquiry and consulting service.

CEH is oriented primarily to United States developments, although significant worldwide events are considered.

SRI is scheduled to launch *World Petrochemicals* (formerly *World Hydrocarbons*) in 1978. This will be specifically international in scope and limited to petrochemicals, as indicated by the title. Clients may subscribe to all four or any combination of the segments:

- World aromatics and derivatives
- World ethylene and derivatives
- World propylene and derivatives
- World C_4 hydrocarbons and derivatives

MULTICLIENT STUDIES

Multiclient studies are prepared by consulting organizations. The studies cover in detail the marketing aspects, and often the technology of products or product groups. The basis of most multiclient studies includes numerous field interviews with producers, users and others; study of patents and other literature; and sometimes computer modeling. Since many studies cover topics in-depth, the price of a single study can easily range up to several thousand dollars.

Frequent producers of multiclient studies include:

1 British Sulphur Corp., Ltd., Parnell House, 25 Wilton Rd., London, SW1V 1NH, England (concentrates on world fertilizer industry.)

2 Business Communications Co., 471 Glenbrook Rd., Stamford, CT 06906.

3 Frost and Sullivan, Inc., 106 Fulton St., New York, NY 10038.

4 Charles H. Kline and Co., Inc., 330 Passaic Ave., Fairfield, NJ 07006

5 Arthur D. Little, Inc., 35 Acorn Park, Cambridge, MA 02139.

6 Predicasts, Inc., 200 University Circle Research Center, 11001. Cedar Ave., Cleveland, OH 44106.

7 Peter Sherwood Associates, 60 E. 42nd St., New York, NY 10017.

8 Skeist Laboratories, Inc., 112 Naylon Ave., Livingston, NJ 07039.

9 Springborn Laboratories, Enfield, CT 06082.

10 SRI International, 333 Ravenswood Ave., Menlo Park, CA 94025.

These are examples of the many active organizations. As the reader can expect, they vary widely in size, quality, and emphasis. In the list above, 4, 5, 6, and 10 are probably the largest.

Directories of multiclient studies are available. These include:

1 *Directory of U.S. and Canadian Marketing Surveys and Services.* Charles H. Kline and Co., Inc., 330 Passiac Ave., Fairfield, NJ 07066.

2 *Published Data on European Industrial Markets.* Industrial Aids, Ltd., Terminal House, 52 Grosvenor Gardens, London SW1 WOAU England.

Both directories cited are updated at regular intervals.

PRODUCT DATA

When the manufacturer of a product has been located, the chemist can obtain extensive information about the product by requesting product bulletins and other trade literature from the manufacturer. Some of this material contains information not readily available from any other source, as mentioned earlier. Examples of kinds of information found in trade literature include:

a Specifications
b Physical and chemical properties in detail
d Any handling precautions
d Analytical methods
e Methods of appropriate use in specific applications
f Lists of pertinent articles and other references

Some manufacturers (for example, DuPont) maintain centralized product information centers that can provide considerable data about their products over the telephone.

A convenient source of product data is the *Visual Search Microfilm File* (VSMF), a product of Information Handling Services (14-18). A number of chemicals and other products are included in this collection of product catalogs and specifications in microfilm form.

Several other tools are available to help the chemist and engineer match a material with desired properties. The data given are usually generic rather than for a specific product by a specific manufacturer. Some examples include:

1 *Materials Selector*, published annually by *Materials Engineering*.

2 *Modern Plastics Encyclopedia*, published annually by *Modern Plastics*.

3 *Corrosion Data Survey*. Covers metals and nonmetals. Published by the National Association of Corrosion Engineers, Houston, TX 77001.

4 The several publications of International Plastics Selector, Inc., 2251 San Diego Avenue, San Diego, CA 92110. These match properties with products on a generic basis and also permit access by commercial name and manufacturer.

5 The *Fulmer Materials Optimizer* (14-19). A materials information system for selection and specification of engineering materials. There are four volumes of information on performance and current costs of commercially available metals, plastics, ceramics, and related component manufacturing processes, plus a method of helping select the optimum material for a given application. An updating service is available.

OTHER SOURCES AND TOOLS

An excellent source of chemical marketing and business information on virtually a worldwide basis is Chemical Data Services, Dorset House, Stamford Street, London, SE1 9LU, England. This organization publishes numerous reports about specific products, companies, countries, and other topics. Some examples are listed below.

- *Chemical Plant Data*—Provides information on chemical plants for over 100 basic chemicals worldwide, based on material collated from published sources in many languages. The service is updated by additional information received from producing companies and from other sources. The information is presented in tabular form on a standard plant information sheet giving details of:

  1 Company

  2 Plant location

3 Existing capacity
4 Planned capacity
5 Date of completion and cost
6 Process used and feedstock
7 Licenser
8 Contractor

As new information becomes available, replacement sheets are issued.

- *Chemical Product Data*—Provides, on a replacement basis, published statistics of production, trade, consumption, and capacity for over 100 basic chemicals covering the leading industrial countries, together with lists of producers and other information. The data is reportedly taken from a variety of original sources in many languages, including official statistics, national trade returns, and data from published journals. The service is available in three volumes: Volume 1—Europe, Volume 2—Asia/Africa, Pacific, Volume 3—The Americas. For each product, the following are provided: Product profile—A brief description of the properties, uses and manufacture of the product. Market summary—Statistics of the current market situation, country by country. Market trends—Comparative production, trade, and consumption statistics for previous years. List of major producers and plant locations, country statistics—Separate sheets giving annual statistics of production, trade, and consumption for every country for which these details are available, as well as end-use and trade breakdowns.

- *Chemfact Books*—A series of books published country by country with information on products, plants, and companies. Countries already covered, or scheduled, include these examples: Belgium, the Scandinavian countries, the United Kingdom, and West Germany. Recent additions include France, Italy, the Netherlands, and Spain.

- *The Chemical Industry of Africa and Asia; The Chemical Industry of the Pacific*—These two separately published "Continent Surveys" include such information as profiles on the chemical industry of countries covered, basic information on major chemical companies, producers of major chemicals, and a variety of statistical and directory data. (This has now been replaced by *Chemical Company Profiles: Africa, Asia, Australasia.*)

- *Chemical Company Profiles Western Europe*—Facts and figures for over 1200 companies in 19 Western European nations.

OTHER REMARKS

This is a field that lends itself well to computerization and specifically to on-line access. Much material is already available in this form, and users can expect further, more sophisticated developments.

The original source of much marketing data is the federal government. It is important that users of the data make known to appropriate officials the essentiality of this information. Expressions of interest will help ensure continued availability of what is needed.

15 PROCESS INFORMATION

How individual chemical products are best made in full-scale commercial practice is a matter of utmost interest and importance. Understandably, this information can be difficult or impossible to obtain from the literature. The reason is that efficient manufacture is key to profits and therefore regarded as proprietary.

The chemist or engineer may find the broad features of a process and even considerable detail described in nonproprietary literature. But the fine detail and ongoing improvements which contribute to optimum production efficiency are rarely published in journals and books.

Exceptions to the proprietary approach are organizations such as government agencies (e.g., the Tennessee Valley Authority in its work on fertilizer technology.) These agencies will usually make the fullest possible disclosure and provide virtually any reasonable assistance (information) necessary.

Also, more details are likely to be available for industrial products and processes which are "mature" (older) such as the basic heavy inorganics like sulfuric acid. Other areas in which process detail is likely to be more readily available are those that affect the public good, such as pollution abatement and energy conservation.

Despite the limitations noted, the astute chemist or engineer who knows the technology can sometimes glean valuable insight by careful study of the pertinent literature.

A first step is often to study what is written in the major chemical encyclopedias. These provide any basic background required.

Briefer treatments of process details for the best known

products will be found in such excellent single volume works as *Faith, Keyes and Clark's Industrial Chemicals* (15-1), *Chemical Process Industries* (15-2), and *Riegel's Handbook* (15-3). These are handy for desk use.

Accounts of processes, or related pertinent information, may appear in such journals as *AIChE Journal, Chemical Engineering, Chemical Engineering Progress, Hydrocarbon Processing* (especially the November issue), and *Industrial and Engineering Chemistry* (process design and development edition), *International Chemical Engineering* and other.

Patents are often an invaluable source of information on processes, as explained in Chapter 9. The sources mentioned above, and others, can be accessed directly or via such tools as *Chemical Abstracts* and *Derwent*, both of which are described earlier in this book. But the chemical engineer or chemist will frequently find the most comprehensive detail on process technology and economics for key chemical products in the services described in the following section.

SPECIALIZED SERVICES

Several highly specialized services provide recent process technology and economics information, primarily for important chemicals of commerce, in considerable detail. The evaluations of processes and comparisons of competing processes are significant features.

These sources are relatively expensive (typical costs are more than $12,000 per year), but they can be found in most major chemical companies and in some major universities. The price is well worth it.

Because the information is private to clients of the services, reference to reports from these sources is not ordinarily found in *CA*, unless the information is made publicly available by the source.

Two of the leading services currently available are:

a SRI International, 333 Ravenswood Ave., Menlo Park, CA. 94025—*Process Economics Program (PEP).*

b Chem Systems, Inc., 747 Third Ave., New York, NY 10017—*Process Evaluation and Research Planning Service (PERP).*

The two services cover some of the same products. When this is the case, the chemical engineer or chemist should consult the corresponding reports of both services, because differences in information content, evaluations, and economic estimates may be found.

The SRI *PEP* Service started in 1963 and now totals well over 100 reports. The reports include an intensive technical review and analysis of each basic process. The aim is to establish commercially feasible operating conditions and to assess technical factors underlying present limitations as well as prospects for improvement. Sufficient details are presented to permit verification of design calculations and cost estimates.

Most *PEP* studies relate to industrial chemicals and polymers that are produced in substantial and growing volumes and are experiencing rapid changes in technology. Both well-established and newer products are included in the studies.

The program includes the following components:

- A series of reports on important chemical and refinery products, based on studies of process technology and cost. This is the core of the program.
- *Process Economics Reviews*, a highlighting of implications of certain new developments in the industry (issued every four months).
- A bulletin describing the status of current process studies issued at four-month intervals.
- License monitoring.

- Consultation with members of the program staff on reports and other matters of interest to individual clients.
- An annual *Yearbook* which updates cost figures presented earlier in the reports. In 1977 the scope of the *Yearbook* was expanded to include economics of chemical production in Europe (West Germany) and Japan, as well as the United States (Gulf Coast).

The Chem Systems PERP service is intended to provide:

1 A continuing evaluation of the commercial significance of current technological developments.

2 Realistic planning information on manufacturing economics of key chemicals and petrochemicals.

3 Information on the status of process development efforts of companies throughout the world.

4 Identification of opportunities for purchase or license of partially developed technology.

Both the technological and commercial situations are examined in arriving at forecasts. An analysis of company activities is important, since plant size will be a function of market concentration, as well as engineering and economic factors. Geographic coverage includes the United States, Western Europe, and Japan. The potential commercial impact of technology developed in other geographic regions is considered in any examination of a product area.

The service is designed to be useful in:

1 New project planning.
2 Research and development.
3 Licensing evaluation.
4 Competitor evaluation.
5 Purchasing analysis.
6 Technological forecasting.

The service provides subscribers with these inputs:

1 A series of in-depth reports, each covering the current and future technological, economic, and commercial situation for key chemicals.

2 Four quarterly reports discussing the potential commercial impact of important recent technological developments. Each report develops and presents an economic and commercial analysis of items believed to be significant. These reports represent an effective method of covering many chemicals in some depth during each year, thus enabling clients to build up a process data bank rapidly.

3 A consulting service, whereby subscribers can discuss items of interest with the staff.

4 A program review meeting given by *Chem Systems'* staff for all subscribers at the beginning of each year. This is held twice annually, once in New York for Western Hemisphere and Japanese subscribers and once in Europe for subscribers in or near that region.

REFERENCES

Chapter 1

1. R. E. Maizell, "Information Gathering Patterns and Creativity," *American Documentation.* **11**, 9–17 (January, 1960).

Chapter 5

1. *Chemical Abstracts Service Source Index 1970–1974 Cumulative*, Chemical Abstracts Service, Columbus, OH, 1975. Quarterly supplements.
2. L. I. Callaham, *Russian-English Chemical and Polytechnical Dictionary,* 3rd ed., Wiley, New York, 1975.

   H. H. Neville, N. C. Johnston, and G. V. Boyd, *A New German-English Dictionary for Chemists*, Van Nostrand, London, 1964.

   A. M. Patterson, *French-English Dictionary for Chemists,* 2nd ed., Wiley, New York, 1954.

   A. M. Patterson, *German-English Dictionary for Chemists*, 3rd ed., Wiley, New York, 1950.

   O. Weissbach, *The Beilstein Guide*, Springer-Verlag, Berlin, 1976, pp. 57–92 (English and French equivalents of words most frequently used in *Beilstein*).

Chapter 6

1. O. B. Ramsay, *The Use of Chemical Abstracts,* American Chemical Society, Washington, DC, 1974 (tape-slide combination and workbook). Revised edition may be issued in 1978.
2. *CAS Printed Access Tools*, Chemical Abstracts Service, Columbus, OH, 1977.
3. This statement appears on the table of contents page of each issue of *Chemical Abstracts*.

Chapter 7

1. E. G. Smith and P. A. Baker, *The Wiswesser Line-Formula Chemical Notation (WLN)*, 3rd ed., Chemical Information Management, Cherry Hill, NJ, 1976.

Chapter 8

1. *Directions for Abstractors*, Chemical Abstracts Service, Columbus, OH, 1975.

Chapter 10

1. *Kirk-Othmer Encyclopedia of Chemical Technology*, 2nd ed., Wiley, New York, 1963–1972 (Dates include Supplement and Index Volumes). New edition to issue beginning in 1978.
2. E. Bartholomé et al., Eds., *Ullmanns Encyklopädie der Technischen Chemie*, 4th ed., Verlag Chemie, Weinheim, West Germany, 1972–
3. J. J. McKetta and W. A. Cunningham, Eds., *Encyclopedia of Chemical Processing and Design*, Marcel Dekker, New York, 1976–
4. H. F. Mark, Ed., *Encyclopedia of Polymer Science and Technology*, Wiley, New York, 1964–1972, Supplement, 1976 and 1977.
5. Hans-G. Boit et al., Eds., *Beilsteins Handbuch der organischen Chemie*, 4th ed., 4th Supplement, Springer-Verlag, Berlin, 1972–
6. O. Weissbach, *The Beilstein Guide*, Springer-Verlag, Berlin, 1976. (Available in English).
7. S. Coffey, Ed., *Rodd's Chemistry of Carbon Compounds*, 2nd ed., Elsevier, Amsterdam, 1964–
8. E. Müller, Ed., *Methoden der organischen Chemie (Houben-Weyl)*, 4th ed., Georg Thieme Verlag, Stuttgart, 1958–
9. J. R. A. Pollock and R. Stevens, Eds., *Dictionary of Organic Compounds*, 4th ed., Oxford University Press, London, 1965. Kept up to date by supplements e.g., 13th supplement, J. B. Thompson, Ed., Oxford University Press, London, 1977.

10. *Organic Syntheses,* Wiley, New York, 1941–
11. From *Organic Syntheses,* Vol. 55, p. IX. Reprinted with permission of John Wiley & Sons, Inc.
12. *Organic Reactions,* Wiley, New York, 1942–
13. C. A. Buehler and D. A. Pearson, *Survey of Organic Syntheses,* Vols. 1 and 2, Wiley, New York, 1970 and 1977.
14. I. T. Harrison and S. Harrison, *Compendium of Organic Synthetic Methods,* Vols. 1 and 2, Wiley, New York, 1971 and 1974. Vol. 3 by L. Hegedus published in 1977.
15. Gmelin Institute (M. Becke-Goehring, Director) *Gmelins Handbuch der anorganischen Chemie,* 8th ed., Springer Verlag, Berlin, 1924–
16. J. W. Mellor, *A Comprehensive Treatise on Inorganic and Theoretical Chemistry,* Wiley, New York, 1922–1937 (main volumes), 1956–1972 (supplements).
17. J. C. Bailar, Jr. et al., Eds., *Comprehensive Inorganic Chemistry,* Pergamon, New York, 1973.
18. *Inorganic Syntheses,* McGraw-Hill, New York, 1939–
19. *Books in Print,* R. R. Bowker Co., New York. Published annually.

Chapter 11

1. Some pertinent publications of the U.S. Patent Office include *General Information Concerning Patents; Patents and Inventions: An Information Aid for Inventors;* and *Q & A About Patents.* These are revised frequently and are available from the U.S. Superintendent of Documents, Washington, DC, or at the U.S. Patent Office.
2. *IFI Assignee Index to U.S. Patents* (see p. 148 of text).
3. U.S. Patent Office, *Manual of Classification of Patents,* U.S. Government Printing Office, Washington, DC. Revised periodically. Includes *Alphabetical Index of Subject Matter.*
4. Same as Reference 1.
5. Some of the manuals available from Derwent include *CPI/WPI User Manual; On-Line Instruction Manual;* and others. Revised annually.

6. J. T. Maynard, *Understanding Chemical Patents,* American Chemical Society, Washington, DC. Scheduled for publication in 1978.

7. P. L. Costas, *Introduction to Patents*, American Chemical Society, Washington, DC, 1974. (Audio tape course and manual).

Chapter 12

1. A. R. Churchley, "Sources of Safety Information," *Chemistry and Industry,* No. 13, 524–527, (July 2, 1977).

2. N. V. Steere, Ed., *Safety in the Chemical Laboratory,* Vols. 1–3, Journal of Chemical Education, Springfield, PA, 1967–1974.

3. W. B. Deichmann and H. W. Gerarde, *Toxicology of Drugs and Chemicals,* Academic, New York, 1969.

   R. Gosselin et al., Eds., *Clinical Toxicology of Commercial Products,* 4th ed., Williams and Wilkins, Baltimore, 1976. Online updates of this work may be available through the Chemical Information System noted on p. 80.

   A. Hamilton and H. L. Hardy, *Industrial Toxicology,* 3rd ed., Publishing Sciences Group, 1974.

   G. P. McKinnon and K. Tower, Eds., *Fire Protection Handbook,* 14th ed., National Fire Protection Association, Boston, 1976.

   National Fire Protection Association, *Fire Protection Guide on Hazardous Materials,* 6th ed., Boston, 1975.

   National Safety Council, 444 N. Michigan Ave., Chicago, IL. Publishes a variety of useful materials including Industrial Safety Data Sheets for specific chemicals.

   F. A. Patty, Ed., *Industrial Hygiene and Toxicology,* 2nd ed., Wiley, New York, Vol. 1, 1958, Vol. 2, 1963.

   N. V. Steere, Ed., *Handbook of Laboratory Safety,* 2nd ed., CRC Press, Cleveland, OH, 1971.

4. N. Irving Sax, *Dangerous Properties of Industrial Materials,* 4th ed., New York, Van Nostrand Reinhold, 1975.

5. L. Bretherick, *Handbook of Reactive Chemical Hazards,* CRC Press, Cleveland, OH, 1975. 2nd ed. scheduled for 1978, Butterworths, Woburn, MA.

6. U.S. National Institute for Occupational Safety and Health, *Registry of Toxic Effects of Chemical Substances,* U.S. Government Printing Office, Washington, DC. Published annually.
7. M. Rand, M. Taras, and A. Greenberg, Eds., *Standard Methods for the Examination of Water and Wastewater,* 14th ed., American Public Health Association, Washington, DC, 1975.
8. Morris Katz, Ed., *Methods of Air Sampling and Analysis,* 2nd ed., American Public Health Association, Washington, DC, 1977.
9. *Proceedings of the 31st Industrial Waste Conference,* Purdue University, May 4–6, 1976, Ann Arbor Science Publishers, Ann Arbor, MI, 1977.
10. *Safety in Academic Chemistry Laboratories,* American Chemical Society (Committee on Chemical Safety), Washington, DC, 1976.
11. *Chemical Engineering Progress* Technical Manual 43, *AIChE Pilot Plant Safety Manual,* American Institute of Chemical Engineers, New York, 1972.
12. *Loss Prevention Series,* American Institute of Chemical Engineers, New York. Published approximately annually.
13. *Ammonia Plant Safety Series,* American Institute of Chemical Engineers, New York. Published annually.
14. Manufacturing Chemists Association, *Guide for Safety in the Chemical Laboratory,* 2nd ed., Van Nostrand Reinhold, New York, 1972.
15. National Fire Protection Association, *Manual of Hazardous Chemical Reactions,* 5th ed., Boston, 1975.
16. National Fire Protection Association, *Hazardous Chemical Data,* Boston, 1975.
17. National Fire Protection Association, Publication 45, *Fire Protection for Laboratories Using Chemicals,* 1975.
18. J. E. McKee, Ed., *Water Quality Criteria,* 2nd ed., California State Water Quality Control Board, 1963; U.S. Environmental Protection Agency Water Pollution Control Series, *Water Quality Criteria Data Book,* Vols. 3 and 5 (EPA 18050 GWV05/71 and HLA09/73, U.S. Government Printing Office,

Washington, DC; and annual literature reviews of the *Journal of the Water Pollution Control Federation* (June issue) which include, among many other topics, reviews of aquatic toxicity such as to freshwater fish.

Chapter 13

1. B. J. Zwolinski, et al., Eds., *Comprehensive Index of API 44—TRC Selected Data on Thermodynamics and Spectroscopy*, Publication 100, 2nd ed., Thermodynamics Research Center, Department of Chemistry, Texas A & M University, College Station, TX, 1974.
2. International Council of Scientific Unions Committee on Data for Science and Technology (CODATA), *International Compendium of Numerical Data Projects*, Springer-Verlag, Berlin, 1969. (Currently being revised for a new edition).
3. R. C. Weast, Ed., *Handbook of Chemistry and Physics*, 58th ed., CRC Press, Cleveland, OH, 1977.
4. R. H. Perry and C. H. Chilton, Eds., *Chemical Engineers Handbook*, 5th ed., McGraw-Hill, New York, 1973.
5. A. J. Gordon and R. A. Ford, *The Chemist's Companion*, Wiley, New York, 1972.
6. M. Windholz, Ed., *The Merck Index*, 9th ed., Merck & Co., Rahway, NJ, 1976.
7. *CRC Composite Index for CRC Handbooks*, 2nd ed., CRC Press, Cleveland, OH, 1977.
8. E. W. Washburn, Ed., *International Critical Tables of Numercial Data, Physics, Chemistry & Technology*, Vols. 1–7 and Index, McGraw-Hill, New York, 1926–1933.
9. G. T. Armstrong and R. N. Goldberg, *An Annotated Bibliography of Compiled Thermodynamic Data Sources for Biochemical and Aqueous Systems (1930 to 1975)*, NBS Special Publication 454, U.S. Government Printing Office, Washington, DC, 1976.
10. R. C. Reid, J. M. Prausnitz, and T. K. Sherwood, *The Properties of Gases and Liquids*, 3rd ed., McGraw-Hill, New York, 1977.

Chapter 14

1. N. H. Giragosian, Ed., *Chemical Marketing Research*, Reinhold, New York, 1967.
2. *Directory of Chemical Producers—U.S.A.*, SRI International, Menlo Park, CA. Published annually with Supplements.
3. *Chemical Week Buyer's Guide Issue*, McGraw-Hill, New York. Published annually.
4. *OPD Chemical Buyer's Directory*, Schnell Publishing Co., New York. Published annually.
5. *Chem Sources—U.S.A.*, Directories Publishing Co., Flemington, NJ. Published annually.
6. *Soap Cosmetics Chemical Specialties Blue Book Issue*, McNair-Dorland Co., New York. Published annually.
7. *Lockwood's Directory of the Paper & Allied Trades*, Vance Publishing Corp., New York. Published annually.
8. *Farm Chemicals Handbook*, Meister Publishing Co., Willoughby, OH. Published annually.
9. *Analytical Chemistry LabGuide*, American Chemical Society, Washington, DC. Published annually.
10. C. S. Cronin, Ed., *Chemical Engineering 1977–78 Equipment Buyer's Guide*, 4th ed., McGraw-Hill, New York, 1977.
11. *Japan Chemical Week*, Ed., *Japan Chemical Directory*, The Chemical Daily Co., Ltd., Tokyo, Japan. Published annually.
12. *Chem Sources—Europe*, Directories Publishing Co., Flemington, NJ. Revised edition scheduled for 1978.
13. *Chemistry and Industry Buyers' Guide*, Society of Chemical Industry, London. Annual publication may be anticipated.
14. *European Chemical Buyer's Guide*, Chemical Data Services, London, 1975. New edition scheduled for 1978.
15. *The Worldwide Chemical Directory*, Chemical Data Services, London, 1977.
16. *Thomas Register of Manufacturers & Thomas Register Catalog File*, Thomas Publishing Co., New York. Published annually.
17. *Custom Processing Services Guide*, 2nd ed., Roger Williams Technical and Economic Services, Princeton, NJ, 1976.

18. Information Handling Services is located at 15 Inverness Way East, Englewood, Colorado.
19. Fulmer Research Institute Ltd., Stoke Poges, Slough, Buckinghamshire, U.K. (See description in *Chemical Engineering*, December 23, 1974, p. 42).
20. *Moody's*$_R$ *Industrial Manual*, Moody's Investors Service, New York. Published annually with supplements.
21. *Dun & Bradstreet*[R] *Million Dollar Directory*[R]. Dun & Bradstreet, New York. Published annually.
22. *Standard and Poor's Register of Corporations*, Standard and Poor's, New York. Published annually.

Chapter 15

1. F. A. Lowenheim and M. K. Moran, *Faith, Keyes, and Clark's Industrial Chemicals*, 4th ed., Wiley, New York, 1975.
2. R. N. Shreve and J. A. Brink, Jr., *Chemical Process Industries*, 4th ed., New York, McGraw-Hill, 1977.
3. J. A. Kent, Ed., *Riegel's Handbook of Industrial Chemistry*, 7th ed., New York, Van Nostrand Reinhold, 1974.

APPENDIX

The purposes of this section are to update the text based on a selection of recent developments, to include some newer material not readily available to the author when the original manuscript was prepared, and to clarify or reemphasize as appropriate. As in the case of the full text, this section is not all inclusive but rather is representative of the ever-changing spectrum of chemical information.

Page references given are to the most pertinent sections of the text.

P. 21 In addition to *ASCATOPICS*, customized or tailored *ASCA* service is available from the Institute for Scientific Information and can be of special value when the chemist's area of interest overlaps into the biological or life sciences.

P. 23 As of about late 1978, *CA SELECTS* had expanded to more than 75 topics. Speculation has it that the series will grow still further and that all United Kingdom Chemical Information Service (UKCIS) Macroprofiles will be available directly from *CA*, perhaps beginning in 1979, in a continuation of the cooperative effort.

P. 33 According to some sources, SSIE may be federalized and transferred to the Department of Commerce as part of NTIS. This speculation may be resolved in 1979.

P. 55 Chemists who work with polymers are especially interested in nomenclature questions. They point to the structural uncertainty and interdisciplinary nature which characterizes much polymer research and literature. Various groups are at work to achieve further improvement in polymer nomenclature and indexing.

P. 68 The availability of materials covered by *CA* was thought to be facilitated when Chemical Abstracts Service (CAS) and NTIS implemented an agreement under which a copy of many of the original journal articles cited in CA could be obtained from either NTIS or CAS. But the outlook is now uncertain, with an apparent change in NTIS policy.
P. 80 The Chemical Information System is presently a complex of a number of interlinked information modules operated by Interactive Sciences Corp., Braintree, MA 02184. Examples of the components available for use include:

MSSS—The Mass Spectral Search System

CRYST—Crystallographic Data Search System

CNMR—Carbon-13 Nuclear Magnetic Resonance Search System

SANSS—Structure and Nomenclature Search System

CAMSEQ-II—Conformational Analysis of Molecules in Solution by Empirical and Quantum-Mechanical Techniques

MLAB—Modeling Laboratory

RTECS—Registry of Toxic Effects of Chemical Substances

The data bases have all been registered by the Chemical Abstracts Service, so that the CAS Registry Number can be used to link between data bases.
P. 80 A UK on-line information service is know as *InFoLine,* one of the host systems to be connected to the European service *Euronet.* Derwent is one of the five sponsors. It is expected that American users will be able to access the files through the overseas networks.
P. 110 A second supplement volume, published December 1977, continues the updating effort for *Mark;* periodic supplement volumes are planned for the future.
P. 112 As of 1978, a 34-page booklet, *How to Use Beilstein,* was available free from Springer-Verlag New York Inc., 175 Fifth Avenue, New York, NY 10010. Additionally, an audiovisual guide to *Beilstein* can be purchased from Science Media, P.O. Box 910, Boca Raton, FL 33432. Preparation is

currently in progress for the fifth supplementary series of *Beilstein* which will cover the literature from 1960–1979, while at the same time, work on completing the fourth supplementary series continues. Results later than the nominal closing dates of the series are claimed to be included wherever possible. Reiner Lukenbach became director of the Beilstein Institute for the Literature of Organic Chemistry in 1978.

P. 120 As part of its *CA SELECTS* series (see also p. 23), *CA* has introduced *New Books in Chemistry* which, if continued, should be a useful aid in keeping up with and identifying books of interest to chemists and chemical engineers.

P. 136 In addition to its coverage of patents, Derwent offers services on other kinds of scientific literature in certain fields. These services include *Ringdoc* (pharmaceutical literature from 1964), *Vetdoc* (veterinary literature from 1968), *Pestdoc* (pesticidal literature from 1968), and *CRDS* (chemical reactions patents and literature from 1975). *CRDS* is scheduled to be available on line through System Development Corp. It is based on the *Journal of Synthetic Methods* and provides coverage of new developments in synthetic organic chemistry. Dr. W. Theilheimer, whose yearbook, *Synthetic Methods of Organic Chemistry*, is well known, is a contributor to this effort.

P. 143 Some newer developments in *CA* patent coverage and other points worth reemphasizing include these:

1. As a result of discussions with the United States Patent and Trademark Office, *CA* has determined that all division, continuation, and reissue patents can contain no new technical material. Therefore, since January 1, 1978, all such documents are referenced in the Patent Concordance rather than abstracted as before. Since continuation-in-part documents *can* contain new technical material, *CA* will continue to write abstracts for such United States documents. Since July 1, 1978, CAS has extended the treatment of United States domestically related documents to domestically related documents from several other countries. Division patent documents from Canada, Federal Republic of Germany, Japan, The Netherlands, and South Africa do not contain new technical material and are referenced to their predecessors

or foreign equivalents of their predecessors in the Patent Concordance. All other non-United States domestically related documents covered by CAS may contain new technical material and are abstracted and indexed.

2. *CA* has recently begun publishing a "Markush Structure" which is reflective of the claim portion of the patent.

3. *CA* has begun to designate patents by the capital letter "P" in the *issue keyword index*. This permits users of these indexes to immediately identify patent citations.

4. *CA* treats patent and journal literature similarly. New chemical information is emphasized, and legal information such as proprietary rights is of little concern at this time. *CA* indexes only those compounds that have been reduced to practice, as evidenced by positive indication of preparation, method of manufacture, or use.

The above list is excerpted from a paper presented by Dr. Philip J. Pollick of the CAS staff before a committee meeting of the Pharmaceutical Manufacturers Association on March 23, 1978. Material reproduced from *Chemical Abstracts* is copyrighted by the American Chemical Society and is reproduced with their permission. No further copying is permitted.

For an explanation of some of the terms and concepts noted above, the reader should consult guides to patents such as References 1 and 6 for Chapter 11.

P. 155 Derwent now covers *Research Disclosure* (*RD*), a journal of articles published in Hampshire, England by Industrial Opportunities, Ltd. The pages of *RD* are available to companies who, for certain inventions, seek an alternative or supplement to obtaining patents and may wish both to provide benefit from the prompt publication of technology and to maintain freedom for their own use of such technology. Disclosures published in the journal may be anonymous. Note that *RD* is covered also by *CA* and probably by other abstracting and indexing services, as appropriate.

P. 157 By intent, the text contains no special section devoted to energy-related research and development, although there is some overlap with environmental information. One obvious source of information about the chemical and chemical engineering aspects of energy is *CA*, especially in its sections 50, 51, 52, 69, 70, 71, and 72. The United States Department of Energy (DOE) prepares a comprehensive semimonthly journal *Energy Research Abstracts*. DOE also maintains or monitors other scientific and technical files, including their on-line RECON System. A variety of reputable for-profit organizations provide important energy-related services; contact with the Information Industry Association will help identify some of these.

P. 163 NIOSH is a fountain of technical information related to occupational safety and health. At its offices in Cincinnati, OH 45226, NIOSH has established the Clearinghouse for Occupational Safety and Health Information. Although primarily intended originally for internal NIOSH use, the Clearinghouse can help qualified nongovernment users locate needed data, and there are plans to make its computer-based information storage and retrieval system accessible to the public in an online mode. NIOSH also has several additional specialized files. The results of NIOSH research have been extensively disseminated through the United States Goverment Printing Office and the United States National Technical Information Service.

P. 167 The latest available news about the *Toxicology Data Bank* is that the file contains information on 1100 substances plus information on some additional 1500 which are in process of "data extraction." This means that about 2600 substances are in the first, public file.

P.168 A good recent and concise summary of major United States environmental regulations affecting the chemical industry is found in a paper by N. R. Passow, *Chemical Engineering*, November 20, 1978, pp. 173–180.

P. 169 Early in 1979, chemists and engineers can expect to have available the initial inventory of commercial chemicals published in accordance with the federal Toxic Substances Control Act (TOSCA). This should be the most complete listing of "chemicals of commerce" ever published in the United States; it will probably appear first in the *Federal Register*. Although names of manufacturers will probably not be given, the very listing of such chemicals will be valuable, as will such anticipated supplementary data as synonyms, *CA* Registry Numbers, and other modes of identification.

P. 170 In February, 1978, a government-wide United States Interagency Toxic Substances Data Committee was organized to coordinate the many federal activities and files on chemical information, with special emphasis on toxicology and related properties and on the regulatory aspects. Some other kinds of chemical information are also being considered. The committee is working on the development of a "Chemical Substances Information Network" (CSIN). When fully implemented, such a network should be enormously useful to chemists in and out of government. Completion of the proposed network could take about 10 years, although some parts may be operational in 1979.

P. 173 *The Loss Prevention Bulletin* of the Institution of Chemical Engineers is an information exchange plan under which organizations submit reports of incidents in their plants for editing to ensure anonymity. Also published in the *Bulletin* is technical comment on the reported incidents, in addition to articles and safety information submitted by subscribers. The Institution is in the process of publishing a series of *Hazard Workshop Notes*, which complement the bi-monthly *Bulletin*. The *Notes* are designed to be used as a basis for discussion of safety-related subjects.

P. 177 One convenient way of keeping up with pertinent environmental and related reports is the weekly National Technical Information Service abstract bulletin *Environmental Pollution and Control*. This includes abstracts of government-sponsored and other reports on such topics as environmental health and safety, environmental impact statements, air, noise, solid wastes, water, pesticides, and radiation.

P. 181 In addition to what is in the literature, some firms can assist in scientific evaluation of potential industrial hazards. One example is Hercules, Inc., P.O. Box 210, Cumberland, MD 21502. As part of its services, Hercules publishes data guides and newsletters that emphasize flammability data.

P. 195 A newer activity at the Thermodynamics Research Center, Texas A & M University, is their computer data bank which can be used to generate physical and thermodynamic properties of fluids. This is presently an experimental project. R. C. Wilhoit is the contact for further information.

P. 200 Another useful handbook is *Physical Properties*, prepared under the editorship of C. L. Yaws and published by the magazine *Chemical Engineering* as a book in 1977. The kinds of data covered include heat of vaporization, vapor pressure, heat capacity, surface tension, viscosity, thermal conductivity, and density. Information that can aid in computer-based calculations is included.

P. 211 Beginning with the issue of June 5, 1978 (Volume 85, No. 13) *Chemical Engineering* carries a series of articles by J. N. Peterson et al., "Computer Programs for Chemical Engineers: 1978." This series effectively obsoletes and updates the 1973 compilation. The complete 1978 series is available from *Chemical Engineering* as Reprint No. 296. It was prepared in connection with a project funded by the United States Department of Energy at M.I.T. to develop an advanced software computing system for chemical process engineering (Project ASPEN).

P. 211 At least part of the hopes of "Project Evergreen" appear to have emerged in the form of the Design Institute for Physical Property Data. This new group was established by the American Institute of Chemical Engineers with academic leadership and strong industry support. Goals include the collection and evaluation of more and better data needed for equipment and process design.

P. 225 A newer product of Chemical Data Services is the *IPC Chemical Data Base* for chemical products worldwide, available as an on-line service on an international basis.

PP. 234–241 Some new publications of special interest to this author include those listed below. Space and time do not permit listing all but a few of the newer, more important sources.

- L. de Vries and H. Kolb, *Dictionary of Chemistry and Chemical Engineering*, 2nd ed., New York, Verlag Chemie, 1978 (German/English and English/German).
- K. Weissermel and H. Arpe, *Industrial Organic Chemistry*, New York, Verlag Chemie, 1978 (English translation).
- M. E. Green and A. Turk, *Safety in Handling and Working with Chemicals*, Macmillan, New York, 1978 (intended for students and technicians).
- *German Chemical Engineering* (this new journal, available from Verlag Chemie beginning in 1978, is one good way to keep up with German technology; it is published in English).

P. 237 A 3rd revised edition of *Patty's Industrial Hygiene and Toxicology* has been published. G. D. Clayton and F. E. Clayton are the editors of the new edition of this classic reference work.

INDEX